RAND

Concept-Level Analytical Procedures for Loading Nonprocessing Communication Satellites with Nonantijam Signals

Edward Bedrosian, Gaylord K. Huth

*Prepared for the
United States Air Force
United States Army*

The command and control of modern military forces is becoming increasingly dependent on space assets for a wide variety of functions. Prominent among these are communication satellites, the value of which has been amply demonstrated in recent military operations, notably Operations Desert Shield/Storm. Unfortunately, modern communication satellite systems are very expensive. Given the shrinking military budget and the volatile geopolitical world in which they must be used, it becomes essential to obtain those systems that best serve these uncertain needs and to do so at the least cost.

As part of its research for the Army and the Air Force, RAND is constructing a concept-level modeling tool that is intended to permit evaluation of conceptual military communication satellite systems at a systems level. That is, it considers only basic design parameters. This report is the first in a series devoted to presenting the analytical procedures required in such a computer model and does not discuss the model's implementation. The second in the series is MR-640-AF/A, *Concept-Level Analytical Procedures for Loading Nonprocessing Communication Satellites with Direct-Sequence, Spread-Spectrum Signals,* by Edward Bedrosian and Gaylord K. Huth, 1996.

This analysis has been conducted jointly under two of RAND's federally funded research and development centers (FFRDCs), Project AIR FORCE and the Arroyo Center.

Project AIR FORCE is the FFRDC operated by RAND for the U.S. Air Force. It is the only Air Force FFRDC charged with policy analysis. Its chief mission is to conduct objective and independent research and analysis on enduring issues of policy, management, technology, and resource allocation that will be of concern to the senior leaders and decisionmakers of the Air Force. Project AIR FORCE work is performed under contract F49620-91-C-0003. The research reported in this document was conducted under the C4I/Space Project within the Force Modernization and Employment Program of Project AIR FORCE.

The Arroyo Center is a studies and analysis FFRDC operated by RAND for the U.S. Army. It provides the Army with objective, independent analytic research on major policy and organizational concerns, emphasizing mid- and long-term problems. Arroyo Center work is performed under contract MDA903-91-C-0006. The research reported in this document was conducted under the C3I/Space for Contingency Operations Project within the Force Development and Technology Program of the Arroyo Center.

CONTENTS

FIGURES

This report is the first of a series devoted to presenting the analytical procedures and mathematical formulations required to construct a computer model of a military communication satellite system, load it efficiently with the radio signals required to support an operational scenario, and assess its vulnerability to jamming. It does not address the implementation of the model.

Inasmuch as the model is intended to facilitate relative, rather than absolute, comparisons between various communication satellite systems, both real and conceptual, only the essential technical characteristics of these systems and the terrestrial terminals with which they are intended to operate are considered. This not only simplifies the construction and operation of the model, but it also focuses attention on those elements of the overall system that are of the greatest significance in a comparative analysis.

A discussion of scenarios and their various phases during a military operation are presented in Chapter Two. From this, the communication requirements can be derived in the form of a network diagram and a connectivity matrix. An example is presented to illustrate the procedure. The earth stations typically used with communication satellites are described in general terms in Chapter Three and their essential components are identified. The internal configuration of the earth station serving one of the exemplary network users is then presented to illustrate the data flow that must be accommodated.

The payload configuration of a Defense Satellite Communication System (DSCS) III satellite is presented in Chapter Four as an example of the type of frequency-translating, nonprocessing communication satellites that are to be considered. Two such satellites are used in Chapter Five to illustrate a system configuration consistent with the exemplary network developed in Chapter Two. The signal flow through those two satellites and the other user earth stations for the exemplary network are then detailed to complete the concept-level illustration.

The assumptions to be used about data rates, modulation, guard bands, etc., are presented in Chapter Six for the type of nonantijam signaling of interest in this report. Representative values for certain key parameters are presented for easy reference. The types of data needed in the model database are listed in Chapter Seven. This includes a listing of the principal DSCS terminals and their essential technical characteristics, as well as technical data on the DSCS III satellites.

The formulas required to load a satellite transponder with the signals specified by the network diagram are derived in Chapter Eight. This is done link by link working backward from the receiving terminal through the satellite to the transmitting terminals. The loading proceeds by allocating transponder bandwidth and power link by link until one or the other is completely depleted.

A jamming analysis is presented in Chapter Nine. Although the nonantijam signals being considered are not immune to deliberate interference, it is nonetheless of interest to determine precise vulnerabilities to various types of jamming. Breaklock jamming, in which a circuit is disrupted by disturbing a receiver's phase-locked loop, is shown to be a relatively simple task. Noise jamming is also shown to be effective without great effort. Brute-force continuous-wave jamming is the simplest of all and can disable an entire transponder using a radiated power only 6 dB greater than the sum of all the user-radiated powers. However, such a jammer would not be easily transportable or concealed, even though it would not be as large as many fixed earth stations.

SYMBOLS AND ACRONYMS

A	Generic for effective antenna aperture, m^2
AJ	Antijam
AWGN	Additive White Gaussian Noise
B	Transponder bandwidth, Hz
$(BO)_i$	Input backoff to TWTA or SSPA, dB
$(BO)_o$	Output backoff to TWTA or SSPA, dB
BPSK	Binary Phase Shift Keying
C	Difference in longitude between satellite and user, deg
CW	Continuous Wave
d	Generic subscript denoting downlink
dB	Decibel
DMUX	Demultiplexer
DSCS	Defense Satellite Comunication System
e	Bandwidth factor
E_b	Energy per bit, J
$(E_b/N_0)_{min}$	Threshold signal-to-noise ratio, dB
EC	Earth Coverage

EIRP	Effective Isotropic Radiated Power, dBW
f	Width of guard band as a fraction of the modulated RF signal bandwidth
f_d	Downlink frequency, Hz
FDM	Frequency Division Multiplex
FEC	Forward Error Correction
f_u	Uplink frequency, Hz
G	Generic for antenna gain, dB
h	Altitude of geostationary satellite, 35,784 km
I_0	Modified Bessel function of the first kind of order zero
ICDB	Integrated Communications Data Base
IF	Intermediate Frequency
ISDB	Integrated Satcom Data Base
ITU	International Telecommunications Union
J	Generic for jammer or jammer power
k	Boltzmann's constant, 1.3806×10^{-23} J/K
L	Latitude of user earth terminal, deg
L_i	Receiver input coupling loss, dB
LNA	Low-Noise Amplifier
L_o	Transmitter output coupling loss, dB
M	Generic for margin, dB
MBA	Multibeam Antenna
MHz	Megahertz
MILSATCOM	Military Satellite Communication

MUX	Multiplexers
N_0	Receiver system noise power spectral density, J/Hz
P	Generic for power, dBW
P_e	Output bit error probability
P_i	Input power to TWTA or SSPA, dBW
$P_{lin,0}$	Maximum output power at which TWTA or SSPA is still linear, dBW
$P_{lin,i}$	Maximum input power at which TWTA or SSPA is still linear, dBW
P_o	Output power from TWTA or SSPA, dBW
$P_{sat,i}$	Input power level at which TWTA or SSPA saturates, dBW
$P_{sat,o}$	Output power level at which TWTA or SSPA saturates, dBW
PTT	Post, Telegraph, and Telephone
QPSK	Quaternary Phase Shift Keying
r	Path length between user and satellite, m
R	Code rate
R_b	Information date rate, bps
R_e	Radius of the earth, 6378 km
RF	Radio Frequency
R_{ij}	Data rate desired from user i to user j
$\mathscr{R}$	Generic for power flux density, W/m^2
S	Generic for signal or signal power
SCT	Single-Channel Transponder

SHF	Super High Frequency
SSPA	Solid State Power Amplifier
T	Receiver system noise temperature, K
TG	Transponder gain, dB
TT&C	Telemetry, Tracking, and Command
TWTA	Traveling Wave Tube Amplifier
u	Generic subscript denoting uplink
UHF	Ultra High Frequency
W	Total bandwidth allocated to a given modulated RF signal, Hz
W'	Bandwidth of the coded and filtered modulated RF or IF signal, Hz
ε	Elevation angle of line of sight to satellite with respect to local tangent plane at user, deg
ε_{min}	Minimum allowable value of ε
θ	Geocentric angle between satellite and user, deg
λ	Power ratio of the sinusoidal and Gaussian components at the interference at the input to a hard limiter
λ_d	Downlink wavelength, m
λ_u	Uplink wavelength, m
Λ	Ratio of input to output signal-to-interference power ratios
Σ	Generic for the sum of several signals
ϕ	Off-nadir angle of line of sight from satellite to user, deg

Chapter One

INTRODUCTION

Three main points of view have to be considered for military use of satellite communications: those of the communication user, the satellite communication system operator, and the military planner. From the user's point of view, information needs to be communicated to other users in a timely fashion. Thus, the user is concerned with connectivity to the other users, the reliability of the various links, and the data rate required to provide a timely delivery of the information. From the point of view of the satellite communication system operator, the two prime resources that can be allocated to each satellite in a particular orbit are bandwidth and transmitter power to an antenna coverage area on the earth. The military planner must match the data rate and the reliability and connectivity requirements of the user with the resource capabilities of the various satellite communication system constellations. When using commercial satellite systems, the military planner must also consider ground terminal cost, the cost to use the satellite resources, and the cost charged by the Post, Telegraph, and Telephone (PTT) in each country for ground connections (e.g., landing rights).

To permit the military planner to evaluate both performance and cost of conceptual communication satellite systems for military operational scenarios at a systems level, RAND is developing a concept-level modeling tool. The model considers only basic satellite, earth terminal, and modulator/demodulator (modem) design parameters. This report is the first in a series devoted to presenting the analytical procedures required in such a computer model and does not discuss its implementation.

The analysis presented in this report is restricted to nonantijam signals through nonprocessing communication satellites, i.e., those employing simple frequency-translating transponders. For such satellites, it is sufficient to specify the characteristics of the receiving and transmitting antennas, the channelization system by which the various receiving and transmitting antennas can be interconnected, and the reception and amplification properties of the various transponders. An earth terminal is specified by its transmitter power, antenna gain, and receiver noise temperature. The modem is specified by the signaling format, the data rate, the forward error-correction coding, and the required bit error probability.

To allow comparative analysis, the concept-level modeling tool will be assumed to be driven by the communication requirements associated with a specific scenario. The scenario developed by the military planner specifies the location of the terminals/modems and their characteristics, including data rate, the connectivity between terminals, and the available satellite communication system constellations. Although these scenario communication requirements could be derived from the Integrated Communications Data Base (ICDB), such databases are usually undesirable in this application because they tend to be global in nature, servicewide in scope, and based on broadly defined missions. Thus, they tend to be too general and, often, excessive in nature. Furthermore, they may be incompletely familiar with the capabilities and limitations of satellite communication systems. Scenarios, on the other hand, are developed by experienced military analysts and, therefore, are usually more realistic. When possible, they are based on operational experience, as exemplified by Operations Desert Shield/Storm. Also, being more restricted in scope, scenarios are uniquely capable of representing future contingency operations. This flexibility to reflect contingency operations that range considerably in scope, coupled with the fact that they can be used singly or in pairs (or more, to cover aggravated conditions) makes scenarios capable of exercising conceptual communication satellite systems realistically and fairly.

Chapter Two

SCENARIOS

The analysis to follow assumes the use of geostationary satellites and fixed or transportable (as opposed to mobile) earth terminals. Hence, the coverage area will be fixed during normal operation and Doppler shifts need not be considered. The scenarios will generally be restricted to theaters whose dimensions do not exceed the coverage areas of the communication satellite antennas to be used. For most purposes, this would require restricting theaters to a maximum dimension of 1000 to 2000 km. The analysis also adopts the practice of guaranteeing service at the edge of coverage. This has the advantage of permitting freedom of movement within the area of coverage, albeit at the possible expense of transponder efficiency (in case it is power-limited when fully loaded), because the received signal will be up to 3 or 4 dB greater elsewhere. Hence, the user locations need not be specified precisely. That is, it is enough to know that they will be in the coverage areas of whichever communication satellite antennas cover the theater. Such a restriction should not significantly affect the utility of the model, because even major contingency operations can be accommodated; examples are Southwest Asia and Korea.

Because all military operations proceed in phases, such as crisis, deployment, combat operations, and recovery, scenarios can present "snapshots" taken at various times during an operation. Although the greatest communication requirements will probably occur during the combat operations phase, this may not place the greatest stress on the communication satellite system. This is because of the problems inherent in deploying the earth stations, which may have to be more transportable than those used later, and of the delays in repositioning the communication satellites.

A scenario must identify the various users and the data rates required at a given time in both directions between each pair of users that desire to be interconnected. This amounts to specifying all of the unidirectional, i.e., possibly unsymmetric, comsat communication links that make up the comsat-supported communication networks. Thus, the data-rate specification for a scenario involving N users will consist of an $N \times N$ matrix, as illustrated in Figure 1a. Let R_{ij} denote the data rate desired from user i to user j, with R_{ii}, the diagonal term, being meaningless. If the data rates between users i and j are identical in both directions, $R_{ij} = R_{ji}$ and the matrix is symmetric about the diagonal. The 6×6 matrix shown in Figure 1a, with zeros where appropriate, corresponds to the network diagrammed in Figure 1b. If desired, priorities can be associated with the various links. When not specified, they will all be assumed to have equal priority. Note that the quality of service is not an element in the description of the network connectivity and the link data rates.

The users referred to in the discussion of scenarios are not to be thought of as individuals or specific data sources. Rather, the scenarios will identify the users as specific military units or command elements. The data rates associated with the various users will then be the aggregates obtained by adding together the requirements of their composite elements. In fact, the user will generally have widespread elements that are joined together with terrestrial links. As illustrated in Figure 1, the data rates are aggregated according to the users to and from which they are intended. Thus, user 1 is shown to require a data rate R_{12} to user 2, R_{13} to user 3, and R_{14} to user 4, with the sum $R_{12} + R_{13} + R_{14}$ representing the total data rate from user 1.

		To					
		1	2	3	4	5	6
	1	—	R_{12}	R_{13}	R_{14}	0	0
	2	R_{21}	—	R_{23}	0	0	R_{26}
From	3	R_{31}	R_{32}	—	0	R_{35}	0
	4	R_{41}	0	0	—	0	R_{46}
	5	0	0	R_{53}	0	—	0
	6	0	R_{62}	0	R_{64}	0	—

a. Connectivity matrix

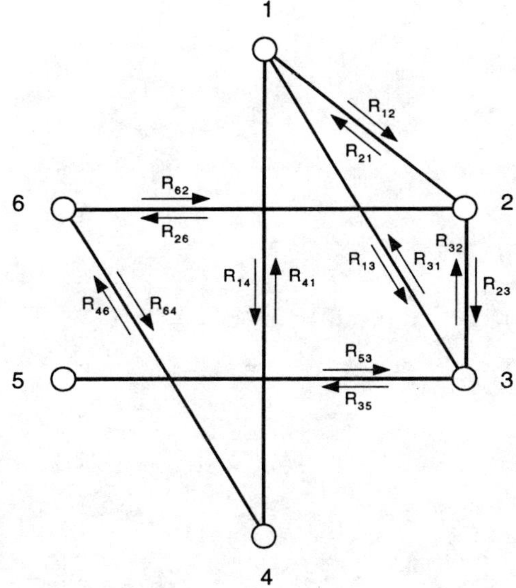

b. Network diagram

Figure 1—Exemplary Communication Network

Chapter Three

THE EARTH STATIONS

An earth station is defined as the assemblage of an earth terminal and one or more modems and switches.[1] The switches are the devices that interconnect the user pairs on the data-stream level. Thus, in the previous example, user 1 has three switches that serve to interconnect it at data rates R_{12} and R_{21} with user 2, R_{13} and R_{31} with user 3, and R_{14} and R_{41} with user 4. Note that the system is transparent to the user above the switch level. The composite data stream from a switch is often referred to as a baseband signal.

Each switch is connected to a modem to which it passes the baseband signal intended for transmission and from which it gets the received baseband signal. The modem is a device that modulates the baseband signal onto an IF (intermediate frequency) carrier for transmission and demodulates the incoming IF carriers on reception. Typical values for the IF are 70 and 700 MHz.

Each earth terminal consists of a (usually) steerable directive antenna that is used with a single specified satellite for both transmission and reception. Each earth terminal also has a wideband amplifier/receiver set that is connected to one or more of the modems. For transmission, the outgoing modulated IF signals from the contributing modems are converted in frequency division multiplex to an appropriate radio frequency (RF), amplified to a specified power level, and sent to the antenna for uplink propagation to the commu-

---

[1]The term switches, rather than multiplexers and demultiplexers, is used here because the context depicted in Figure 2 displays data flowing in both directions. Hence, neither precise term is appropriate alone.

nication satellite. On reception, the downlink RF signals from the satellite are received by the antenna, amplified to a suitable level, converted to IF demultiplexed, and sent to the proper modems for demodulation to baseband. For the SHF components of MIL-SATCOM, the frequency allocation for the uplink extends from 7900 to 8400 MHz and that for the downlink from 7250 to 7750 MHz.

The entire assemblage is diagrammed in Figure 2 for user 1 on the assumption that it has two earth stations, one of which is using one satellite to interconnect with users 2 and 3, the other using another satellite to interconnect with user 4.

The basic scenario used to construct Figure 2 specifies only interuser data rates. The depiction of user 1 as consisting of eight components is purely illustrative and has no bearing on how the model is to be used. The assumption that user 1 needs two earth terminals, however, is significant and must be determined by the operator of the model by considering the disposition of all of the users and the

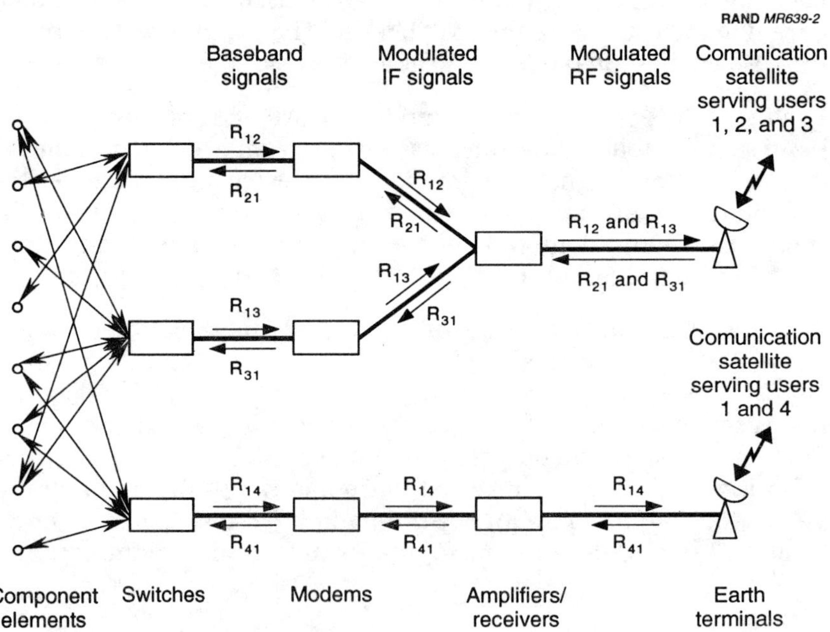

Figure 2—Exemplary Configuration for User 1

available communication satellites. For example, had users 1, 2, 3, and 4 all been served by one communication satellite, a single earth terminal would have sufficed. The situation depicted in Figure 2 might correspond to a case in which users 1, 2, and 3 are all in-theater and can be served by one communication satellite, whereas user 4 might be in the continental United States (CONUS) and can be reached only by a second communication satellite. It can be seen that the scenario must be complete enough with respect to user locations to permit the model operator to determine the required number of earth terminals by considering the types and locations of available communication satellites. If the scenarist does not specify the types of earth terminals available to the various users, the model operator must make suitable selection.

THE COMMUNICATION SATELLITES

Though conceptually simple, communication satellites using frequency-translating transponders often have complex communication subsystem architectures. The following description, based on the DSCS III communication satellite, typifies such communication satellites.

Two major components of the communication subsystem architecture are the frequency plan and the payload configuration. The first of these is shown in Figure 3 and consists basically of a diagram showing the uplink (i.e., receiving) and downlink (i.e., transmitting) frequency bands (or channels) occupied by the various transponders.[1] Guard bands are located between the channels to prevent mutual interference between adjacent channels; beacons are usually located within or at the edges of these guard bands.[2] It is interesting to note that 95 MHz of the total 500 MHz that is allocated at X-band (i.e., 19 percent) is used for intertransponder guard bands in DSCS III. For convenient reference, the saturated power outputs for each channel and the types of receiving and transmitting antennas that are available are also shown.

The connectivities available between various receiving antennas, transponders (i.e., channels), and transmitting antennas are shown in the payload configuration, presented in Figure 4. The essential

---

[1] The frequency plan shown in Figure 3 is typical of the earlier versions of DSCS III, some of which are still in use. Later versions simply downshift all channels by 625 MHz.

[2] The beacons are used for frequency tracking.

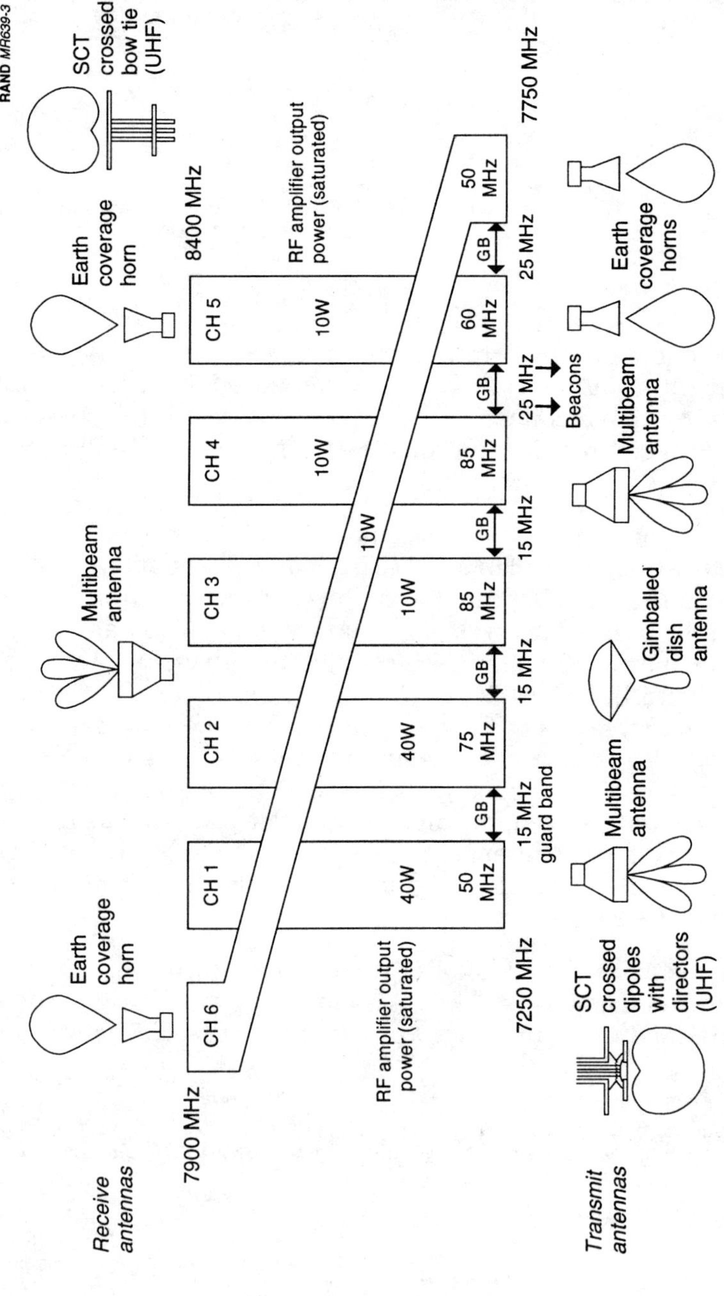

SOURCE: GE Astro Space brochure, n.d.

Figure 3—DSCS III Frequency Plan

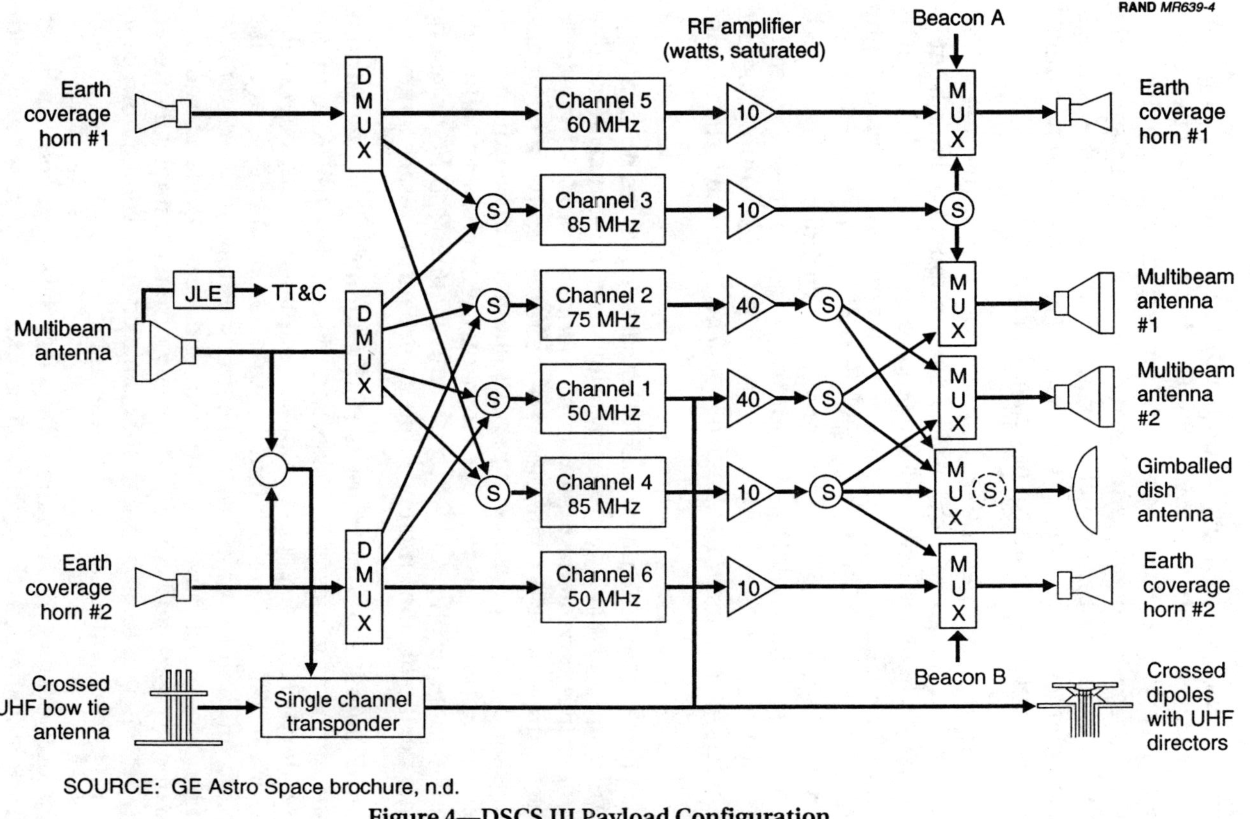

Figure 4—DSCS III Payload Configuration

elements used to arrange a particular desired configuration are the switches (each symbolized by an S). The three super high frequency (SHF) receiving antennas are each connected to a demultiplexer (DMUX) whose function is to separate (i.e., filter, in this case) the uplink signals into the frequency bands appropriate to the channels for which they are intended. The switches are then used to connect the demultiplexer outputs to the desired channels.

Consider, for example, the earth-coverage horn in the upper-left-hand corner of Figures 3 and 4. The three outputs available from its associated demultiplexer indicate that only those uplink signals intended for channels 3, 4, and/or 5 are intended to be received by this antenna. Similarly, the multibeam receiving antenna immediately beneath it is intended to receive only those uplink signals intended for channels 1, 2, 3, and/or 4, whereas the other earth-coverage horn is intended to receive only those uplink signals intended for channels 1, 2, and/or 6. (The UHF subsystem is not considered here.)

It can be seen that the upper and lower earth-coverage horns always drive channels 5 and 6, respectively. However, the four switches that connect the three demultiplexers to channels 1 and 4 provide 16 ways to drive these 4 channels from the three receiving antennas. For example, the upper switch permits channel 3 to be driven by either the upper earth-coverage horn or the multibeam antenna. Similarly, the next lower switch permits channel 2 to be driven by either the multibeam antenna or the lower earth-coverage antenna.

The RF amplifier outputs are similarly connected to the various transmitting antennas by switches and multiplexers (MUXs). In this case, the switches direct the amplifier outputs to the desired multiplexers, where they are combined with other amplifier outputs and passed on to the appropriate transmitting antennas. For example, the output from channel 5 always goes to the earth-coverage horn in the upper-right-hand corner. The output from channel 3, however, can be switched to go either to that same earth-coverage horn or to the multibeam antenna directly beneath it.

It is not likely that the scenario will specify the satellite(s) to be used in the simulation. In terms of existing satellites, the geographical locations specified in the scenario will reveal to the operator of the simulation which are suitable for use because of their orbital locations. Of course, existing satellites can be moved, if desired or necessary, and conceptual satellites can be introduced as desired.

Chapter Five

THE SYSTEM CONFIGURATION

Once the connectivity matrix and diagram are specified, and the earth stations and communication satellites chosen, it is necessary to configure the overall communication system. This is done in Figure 5 for the scenario illustrated in Figure 1 on the assumption that user 4 is located in CONUS, users 1, 2, 3, and 5 are clustered in a relatively small, distant theater of operation, and user 6 is at sea in a nonfixed mid-location. This leads to the need for two communication satellites, which are assumed to be DSCS IIIs. The left-hand DSCS III (#1) is assumed to use earth-coverage horns so that it can span users 1, 4, and 6. The right-hand DSCS III (#2) is assumed to use earth-coverage horns to serve user 6. The users in the theater of operations are assumed to be interconnected using multibeam antennas. The specific payload configurations required for these two satellites are shown in Figures 6 and 7. Note that choices may be considerably restricted in practice where satellite payloads are already configured.

The configuration of user 1 has already been shown in Figure 2. User 6 also requires two earth terminals, but its internal connections are simpler, as diagrammed in Figure 8a. Users 2, 3, 4, and 5 employ single earth terminals but have differing numbers of switches and modems. The configuration for user 4 requires two modems and switches, as shown in Figure 8b. The configuration for user 5 requires only a single switch and modem, as shown in Figure 9a. The configurations for users 2 and 3 are identical, consisting of three switches and modems. This configuration is shown in Figure 9b for the data rates of user 2. For user 3, the appropriate data rates are $R_{31}, R_{13}, R_{32}, R_{23}, R_{35}$, and R_{53}.

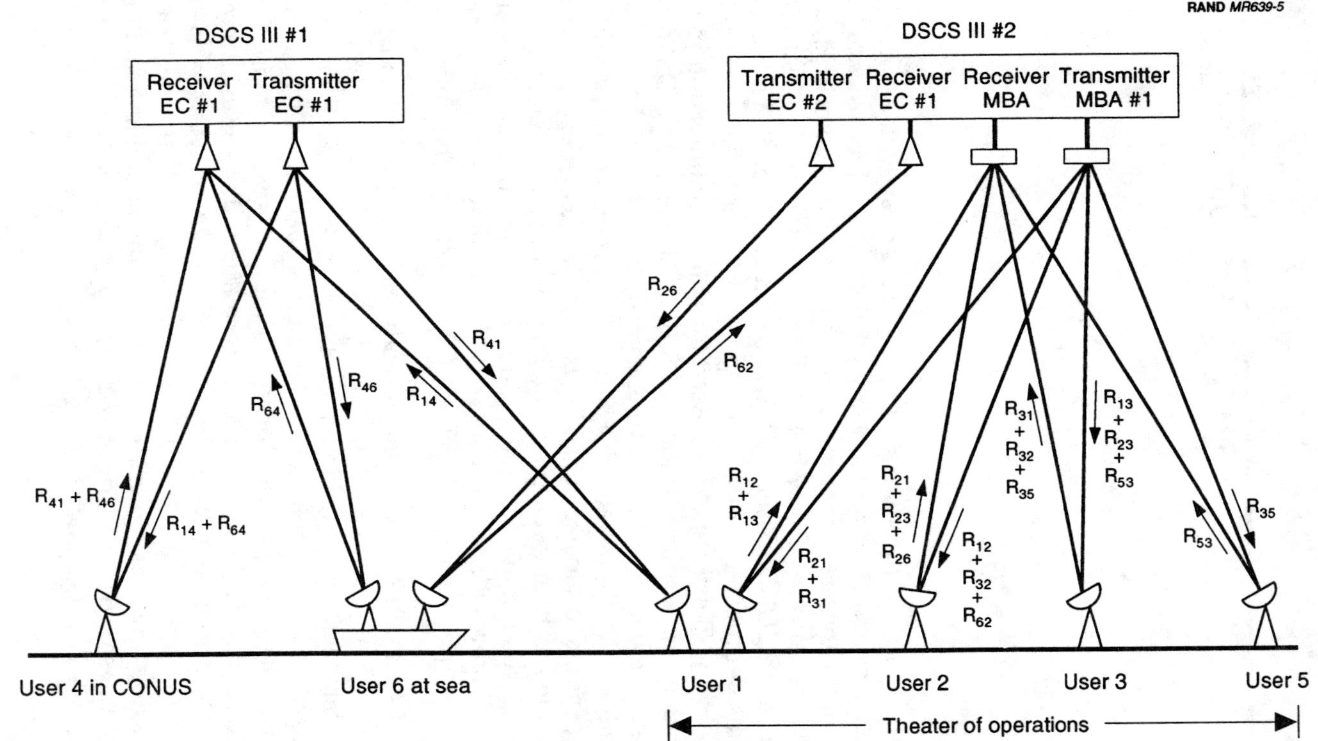

Figure 5—Exemplary System Configuration

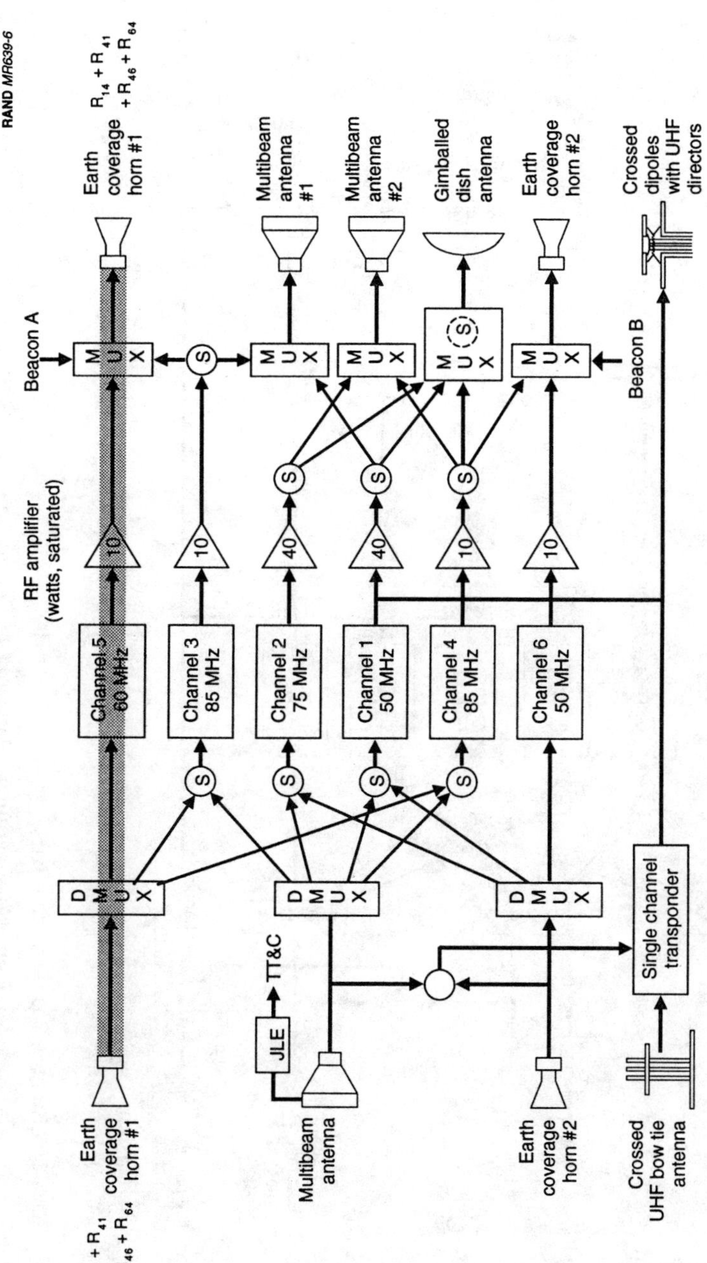

Figure 6—Exemplary Payload Configuration for DSCS III #1

SOURCE: GE Astro Space brochure, n.d.

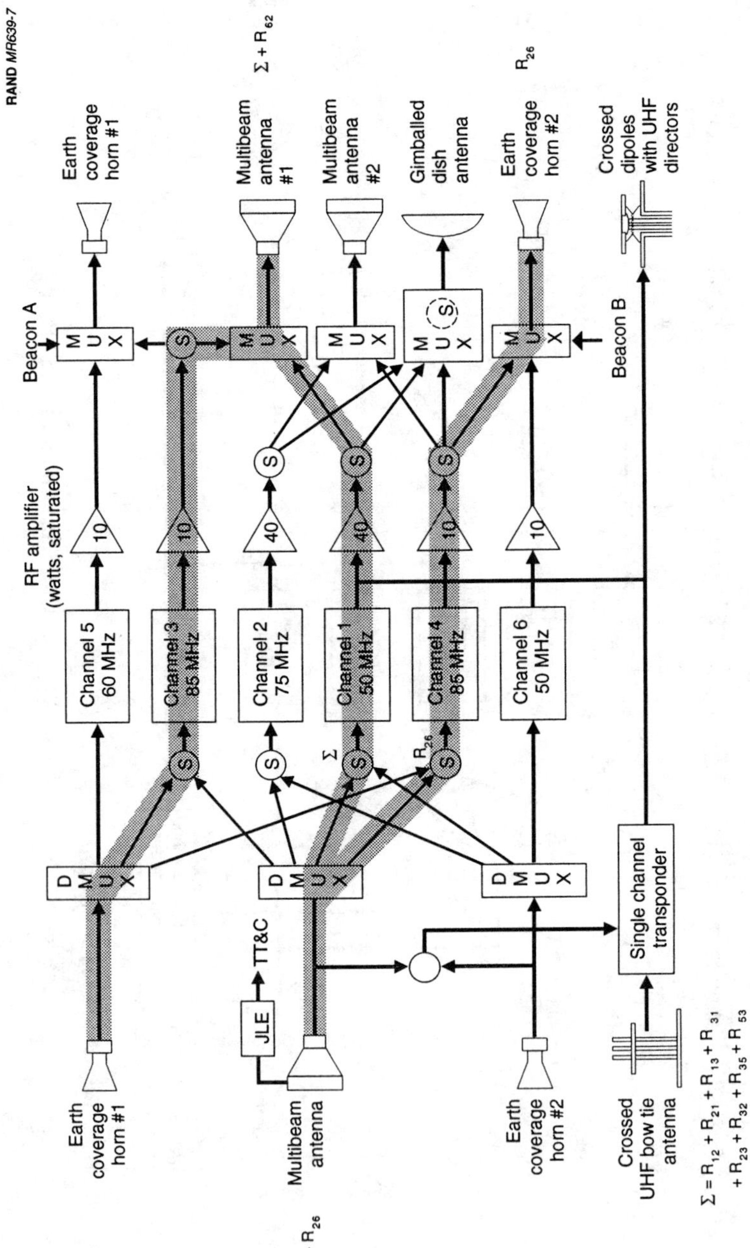

SOURCE: GE Astro Space brochure, n.d.

Figure 7—Exemplary Payload Configuration for DSCS III #2

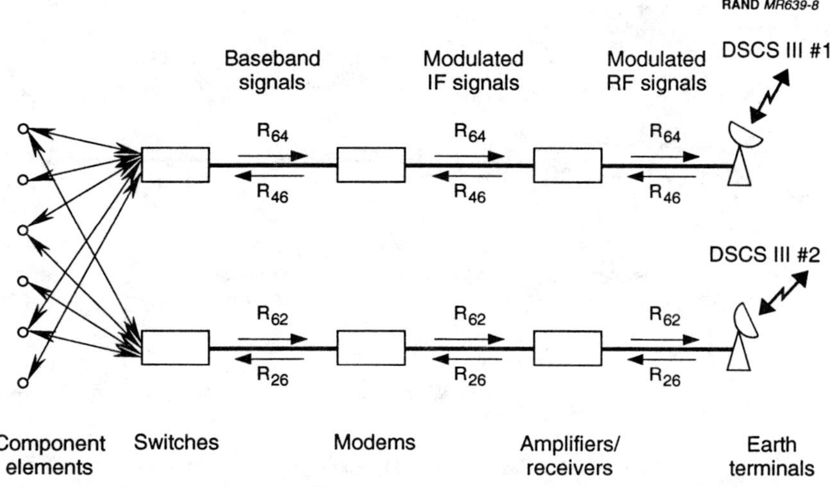

a. User 6

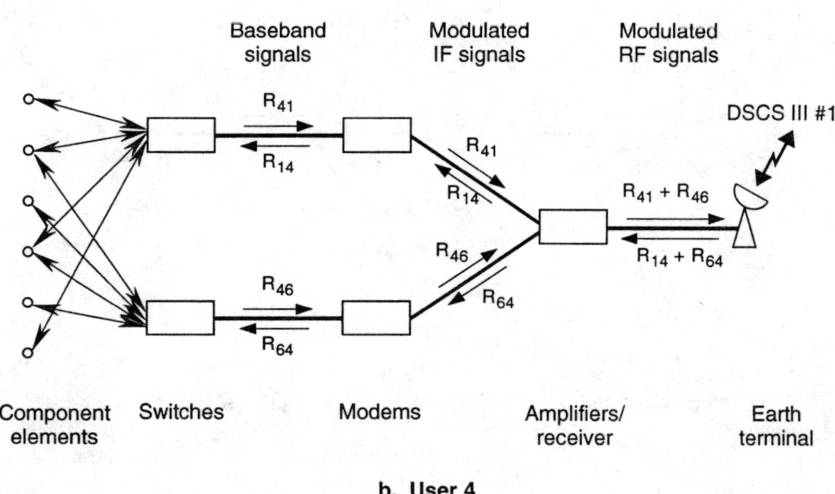

b. User 4

Figure 8—Exemplary Configurations for Users 4 and 6

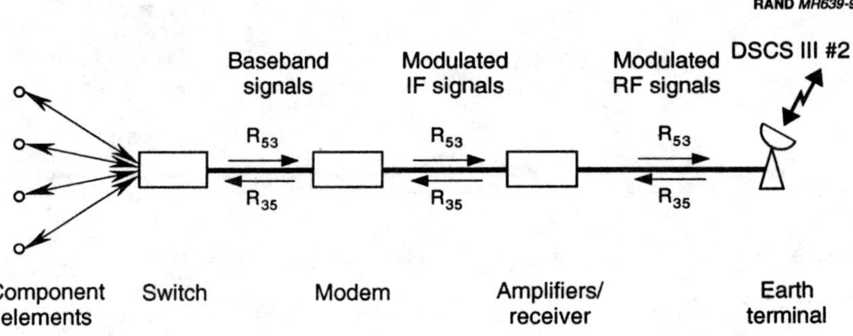

a. User 5

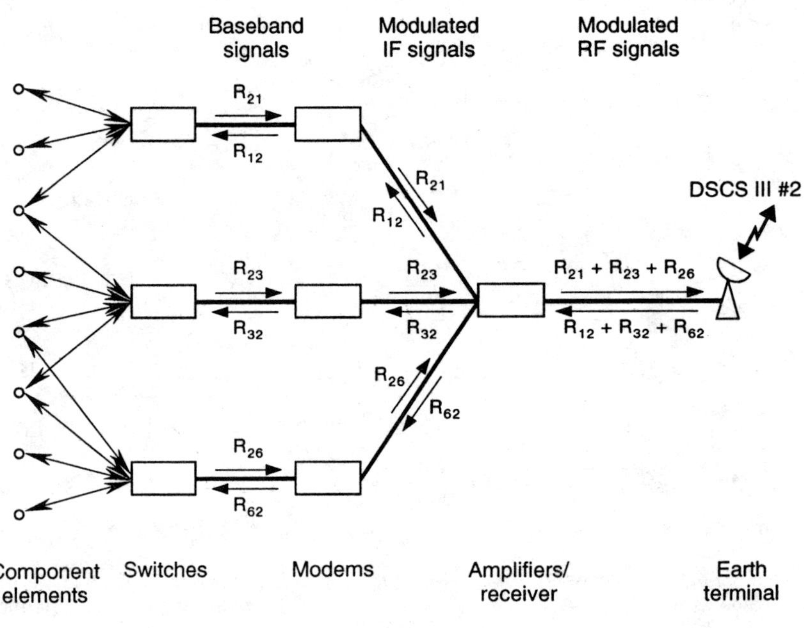

b. User 2

Figure 9—Exemplary Configurations for Users 2 and 5

Chapter Six

ANALYTICAL ASSUMPTIONS

In this chapter, a number of basic system parameters are identified and quantified to the extent necessary to include them in the computer model. Simplifying approximations and suitable formulations are presented as required.

DATA RATES

The data rates specified in the scenario may or may not represent aggregate values that correspond to the capabilities of specific switches and modems. It will be assumed that any of the typical practical data rates may be specified and that the switches and modems can operate at those rates.

MODULATION

Only non-AJ (antijam) modulation techniques will be considered at this time. The choices will be limited to BPSK (binary phase shift keying) and QPSK (quaternary phase shift keying) with or without FEC (forward error correction) coding because these are the ones of greatest interest for the SHF application. Let R_b denote the baseband information data rate and R the code rate ($R=1$ corresponds to no coding). The bandwidth W' of the coded and filtered modulated IF or RF signal is given by

$$W' = e\left(\frac{R_b}{R}\right) \quad e = \begin{cases} 1 & \text{QPSK} \\ 2 & \text{BPSK} \end{cases} \tag{1}$$

where e is a bandwidth factor that depends on the type of modulation to be used. Typical code rates are 1/2, 3/4, 7/8, and 1.

GUARD BANDS

To provide a (minimal) protection against mutual interference between adjacent modulated signals within a given transponder, each modulated signal will be provided with a guard band on either side of the bandwidth W' given by Eq. (1). Let f denote the fractional value of the guard band.[1] Then, the total bandwidth allocation W will be given by

$$W = (1+2f)W' \tag{2}$$

values for f will be solicited, with $f = 0.1$ being the default value.

COUPLING LOSSES

All practical receivers and transmitters exhibit input and output coupling losses L_i and L_o. Typical default values for both are 2 dB at C- and X-bands and 3 dB at K_u-band.

POWER CONTROL

It is assumed that each earth terminal will be assigned and will operate at a precalculated EIRP (effective isotropic radiated power) subject to a suitable maximum EIRP constraint. The earth terminals may exhibit small pointing errors with no error constituting an acceptable default value.

OPERATING FREQUENCY

In practice, loading a real system will require precise knowledge of the operating frequency (or wavelength) of every uplink and downlink signal to prevent mutual interference. For the comparative purposes of calculating the propagation factors in the model, however,

---

[1]A more sophisticated analysis would take into account restrictions on the center frequencies of the signals to avoid objectionable intermodulation products.

such precision is not required and all signals in a given band will be assumed to operate at the band center frequency. The associated uplink and downlink frequencies f_u and f_d and wavelengths λ_u and λ_d are listed in Table 1 for each band using the appropriate band-edge frequencies.

If desired, the center frequencies of the actual up and down bands corresponding to a particular transponder can be used. These bands are indicated for DSCS III in the frequency plan shown in Figure 3.

PATH LENGTH

Initially, the model will deal only with geostationary communication satellites. To calculate the path length to use in the propagation formulas, consider the diagram shown in Figure 10, where the x axis coincides with the earth's polar axis. The y – z plane then contains the equator. Let the z axis contain the satellite S at an altitude h above the surface of the earth, where h = 35,784 km. The radius of the earth is $R_e = 6378$ km.

Consider a user at point P with latitude L (north or south) and difference in longitude C from the subsatellite point O. Let r denote the path length from the user to the satellite, ε the elevation

Table 1

Communication Satellite Band Center Frequencies

C-band	f_u	=	(6.425 + 5.925)/2	=	6.175 GHz,	λ_u =	4.858 cm
	f_d	=	(4.200 + 3.700)/2	=	3.950 GHz,	λ_d =	7.595 cm
X-band[a]	f_u	=	(7.900 + 8.400)/2	=	8.150 GHz,	λ_u =	3.681 cm
	f_d	=	(7.250 + 7.750)/2	=	7.500 GHz,	λ_d =	4.000 cm
K_u-band	f_u	=	(14.000 + 14.500)/2	=	14.250 GHz,	λ_u =	2.105 cm
	f_d	=	(10.950 + 11.700)/2	=	11.325 GHz,	λ_d =	2.649 cm

[a]This band is allocated globally to the fixed-satellite service (i.e., from and to fixed terminals) by the International Telecommunications Union (ITU). (See U.S. Department of Commerce, 1989.) Mobile-satellite service is permitted globally (see footnote 812) in the bands 7.25–7.375 GHz and 7.9–8.025 GHz by special agreement. However, the United States permits military mobile-satellite service in the entire band (7.25–7.75 GHz and 7.9–8.4 GHz) (see footnote G117).

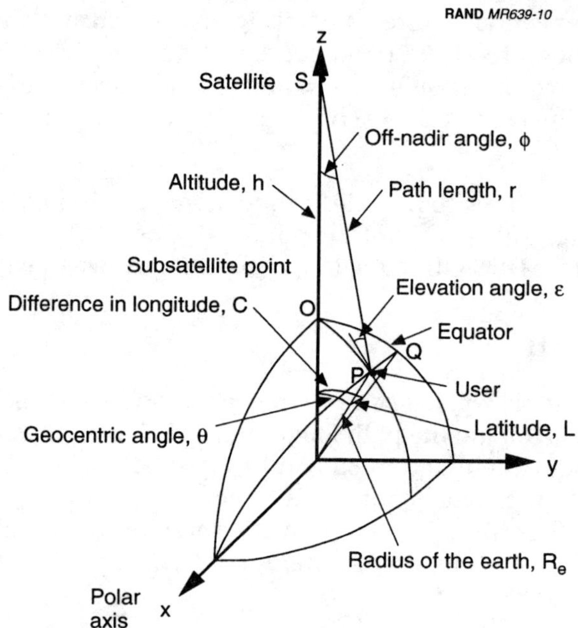

Figure 10—Path Length Geometry

angle of the path at the user, θ the geocentric angle between the satellite and the user, and ϕ the off-nadir angle of the path at the satellite. The elevation angle ε is the angle between the local tangent plane and the line of sight to the satellite. Let ε_{min} denote the minimum allowable elevation angle.

If the operator of the model wishes to calculate the exact path length using the actual satellite and user locations, the following formulas may be used commencing with L and C. The geocentric angle θ as determined by the right spherical triangle OPQ is given by

$$\cos \theta = \cos L \cos C \qquad (3)$$

Then, the path length r becomes

$$r^2 = (h + R_e)^2 + R_e^2 - 2(h + R_e)R_e \cos \theta \qquad (4)$$

To verify that the path results in an acceptable elevation angle ε, note that

$$(h + R_e)^2 = r^2 + R_e^2 - 2rR_e \cos\left(\frac{\pi}{2} + \varepsilon\right) \tag{5}$$

or

$$\sin\varepsilon = \frac{(h + R_e)^2 - r^2 - R_e^2}{2rR_e}$$

where we require

$$\varepsilon \geq \varepsilon_{min} \tag{6}$$

A typical default value for ε_{min} is 10 degrees.

If the operator of the model is content to determine the path length by simply specifying an elevation angle, the off-nadir point angle ϕ is first determined using

$$\frac{h + R_e}{\sin\left(\frac{\pi}{2} + \varepsilon\right)} = \frac{R_e}{\sin\phi}$$

or

$$\sin\phi = \frac{R_e}{h + R_e} \cos\varepsilon \tag{7}$$

The geocentric angle θ is then given by

$$\theta = \pi - \phi - \left(\frac{\pi}{2} + \varepsilon\right) = \frac{\pi}{2} - \phi - \varepsilon \tag{8}$$

and the path length is given again by Eq. (4).

If the default value $\varepsilon = 30$ degrees is used, Eqs. (7), (8), and (4) yield a default path length $r = 38{,}610$ km.

UPLINK NOISE CONTROL

Inasmuch as the satellite simply retransmits on the downlink whatever it receives on the uplink, a precise analysis should include a calculation of the noise contribution by the satellite antenna and receiver front end. However, for simplicity, the model will instead merely verify that the ratio C/N_0 (where C denotes the carrier, or signal, power and N_0 is the effective system noise power spectral density) at the input to the satellite receiver is at least 10 dB greater than the same ratio at the input to the earth terminal receiver. Such an excess will insure that the satellite noise contribution is negligible in comparison with the noise contribution from the earth-terminal receiver. Details of the required formulation appear in Chapter Eight.

TRANSPONDER AMPLIFIERS

Modern frequency-translating communication satellites use traveling wave tube amplifiers (TWTAs) or solid state power amplifiers (SSPAs). Both have input-output power characteristics similar to the one shown in Figure 11. Operation is essentially linear for input power levels P_i equal to or less than a level $P_{lin,i}$. At higher levels, operation becomes increasingly nonlinear until at an input level equal to $P_{sat,i}$, the amplifier is saturated and is producing its maximum or saturated power output. It is customary to relate the input power to the saturation level by referring to an input backoff $(BO)_i$. Corresponding terms with subscript o define the output quantities.

The amount of backoff to use in a specific case depends on the type of amplifier being used and the number and nature of the signals being amplified. For BPSK and QPSK signals, which are the only types considered here, the fact that they have constant-amplitude envelopes (the information being contained in the phase modulation) means that if there is only a single signal, or carrier, being amplified, no backoff is required. That is, the amplifier can be operated at its saturation level.

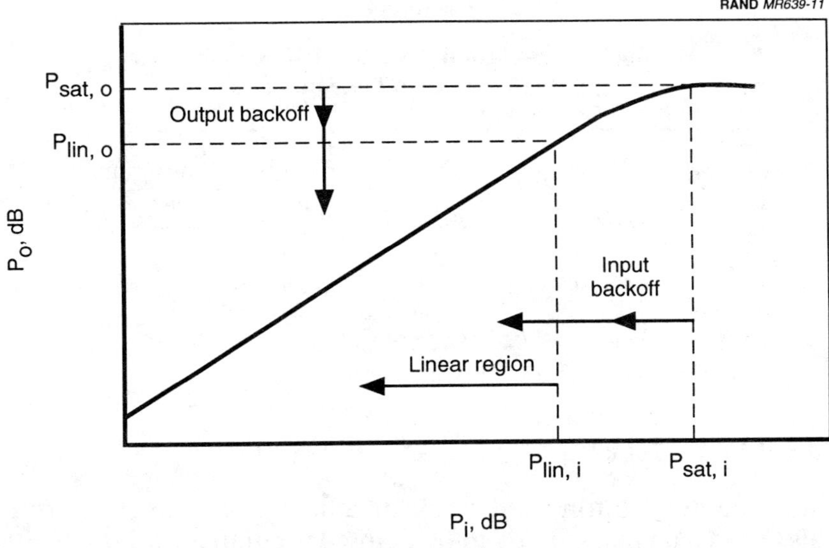

Figure 11—Typical TWTA or SSPA Input-Output Characteristics

When there are two or more carriers, the composite signal is no longer of constant amplitude and harmful intermodulation products will be generated if the amplitude of the composite signal makes excessive excursions into the nonlinear region. Suitable values of input and output backoffs are listed in Table 2. These values are empirically derived and are typical of current technology.

The backoff levels listed for the two-carrier case in Table 2 closely approximate the values needed to operate at the top of the linear region of operation. Within the linear region, it is possible to define a transponder gain (TG) given by

$$(TG) = P_o/P_i \quad P_i \le P_{lin,i} \tag{9}$$

The transponder gain can usually be adjusted stepwise by introducing various amounts of attenuation. The transponder gain does not include the input and output coupling losses but may have slightly different values for each of the transmitting antennas to which it is coupled (see Table 6 in Chapter Seven).

Table 2

**Backoff Levels of BPSK and QPSK Signals
on TWTA and SSPA Transponders**

No. of	TWTA		SSPA	
Carriers	$(BO)_i$,dB	$(BO)_o$,dB	$(BO)_i$,dB	$(BO)_o$,dB
1	0	0	0	0
2	6.0[a]	2.5[a]	4.5[a]	2.0[a]
3	7.5	4.0	6.0	3.5
4	8.5	5.0	7.0	4.5
5 or more	9.5	6.0	8.0	5.5

SOURCE: Adapted from *INTELSAT* (1990).

[a]Also corresponds to backoff to top of linear range.

SATELLITE ANTENNA GAINS AND COVERAGES

Unlike the transmitting and receiving antennas at the earth terminals, which can be assumed to be pointed accurately at the satellite
and which can therefore be credited with their stated gains, the
satellite transmitting and receiving antennas are better specified by
coverage contours. The actual antenna gains are usually realized at
the point on the earth to which the axis of the antenna is directed (or
can be said to be directed in the case of a shaped beam). At other
points within the receiving or transmitting coverage areas, the corresponding received or transmitted signal strength is reduced according to the directivity pattern of the antenna.

The edge of coverage is usually taken as the contour on the earth at
which the directivity is down 3 dB from its maximum value. (Though
it is, strictly speaking, a misuse of the term, satellite receiving and
transmitting antenna gains are usually specified by the "gain" realized at the edge of coverage.) A fixed, i.e., nonrelocatable, user can
calculate and use the directivity for its particular location. Tactical
users, on the other hand, may be relocatable or even mobile, in
which case it may be impractical to attempt such precision.
Inasmuch as the model is intended for use with tactical users, calculations will be made at the edge of coverage using edge-of-coverage
values for the satellite receiving and transmitting antenna gains.

LINK MARGINS

Uplink and downlink propagation paths to a satellite are subject to rain loss, fading, and scintillation. To accommodate these factors, uplink and downlink margins M_u and M_d are included in the link calculations. Typical equal default values are 3 dB at C-band, 4 dB at X-band, and 7 dB at K_u-band.

EARTH-TERMINAL AMPLIFIERS

The amplifiers used in low-power earth terminals may be similar to those found in communication satellites. At the higher power levels, they are likely to be klystrons or similar devices. As a result, although their saturation characteristics may resemble the one shown in Figure 11, there may be different levels of backoff required to operate in the linear region (though great differences are not likely). Consequently, unless specific information is available for a particular earth terminal being considered, Table 2's values will be assumed to apply.

RECEIVER THRESHOLD VALUES

The modem at the receiving earth terminal requires a signal-to-noise ratio that depends, for the BPSK or QPSK systems considered here, on the code rate R that is used and the output bit error probability P_e that is desired. This is shown in Table 3 in terms of a "threshold" $(E_b/N_0)_{min}$ for a bit error probability $P_e = 10^{-6}$. Here, E_b is the received energy per information bit and N_0 is the noise power spectral density. These theoretical values were obtained from Clark and Cain (1981) for rates 1/2 and 3/4, from Haccoun and Bégin (1989) for rate 7/8, and from Lindsey and Simon (1973) for rate 1.

Table 3

**Threshold Values for Coded Systems
at 10^{-6} Bit Error Probability**

Code Rate, R	$(E_b/N_0)_{min}$	
	Theoretical	Practical[a]
1/2	4.9 dB	6.4
3/4	5.7 dB	7.2
7/8	6.7 dB	8.2
1	10.5 dB	12.0

[a]Includes a default value implementation loss of 1.5 dB.

MODEL DATABASE

The earth terminals are adequately specified by their operating band(s), instantaneous bandwidth, maximum EIRP, $EIRP_{max}$, and sensitivity G/T. The maximum allowable data rate per carrier may be desirable. The amplifier maximum power output and the antenna size are useful for reference purposes. This information should be stored in the model database for all existing and planned military and commercial earth terminals. Such a list is shown in Table 4 for the current DSCS terminals.

The communication satellites are adequately specified by their frequency plans (Figure 3, for example), payload configurations (Figure 4, for example), receiving and transmitting coverage diagrams, orbital locations, and transponder technical characteristics. These can be stored in the model database in detail for analysis by the operator as was done in Figures 5 and 6. However, a concise tabulation of the characteristics of the allowable configurations will probably suffice. Such listings are presented in Tables 5 and 6 for DSCS III. The DSCS III transponder gains (TG) are listed in Table 6 for the six transponders in combination with the transmitting antennas to which they can be connected for the 10 gain states that are available.[1] As noted in the previous chapter, the values of transponder gain vary from transponder to transponder according to the transmitting antennas to which they are connected. Note that all antenna gains, receiver sensitivities, and EIRPs are given at the edge of coverage.

---

[1] The reference to a 3 dB (output) backoff in Table 6 confirms, in accordance with the value listed in Table 2 for a TWTA, that operation is at the upper end of the linear region.

Conceptual earth terminals and communication satellites must also be specified in a comparable fashion, ad hoc.

Table 4

DSCS Terminals[a]

Type	Use	Antenna Diameter (ft)	G/T (dB/°K)	EIRP (dBW)	Transmission Power (kW)	Instantaneous Bandwidth (MHz)[b]
FSC-9[c]	Fixed, nodal	60	37	98	12.5	125
MSC-46[c]	Fixed or transportable	40	34	87/93	3/12.5	500/125
TSC-54[c]	Transportable	18	26.5	87	5	50
FSC-78[d]	Fixed, nodal	60	39	97	10	500
GSC-61[d]	Fixed or transportable	38	34	93	10	500
TSC-86 (LT-1)	Transportable	8	18	73	1	50
TSC-86 (LT-2)	Transportable	20	27	81	1	50
TSC-85	Tactical, ground-mobile[e]	8	18	70	0.5	50
TSC-93	Tactical, ground-mobile[e]	8	18	70	0.5	50
TSC-94	Tactical, ground-mobile[e]	8	18	70	0.5	50
TSC-100		8/20	18/27	70/81	1	50
WSC-6	Ship	4	12	65	3	500
SSC-6	Ship	6	15	80	12.5	125
ASC-24[f]	Airborne command post	2.75	7	74	12.5	125

[a]Terminals for diplomatic telecommunications service and special applications are not listed.

[b]Transmission only; all terminals have 500 MHz instantaneous receiver bandwidth.

[c]These are modified Phase I terminals. All others are new developments for Phase II.

[d]The prototype FSC-78 was designated the MSC-60; the MSC-61 development became the GSC-61.

[e]No operation while in motion; setup time at destination is 20 min.

[f]The development phase terminal was designated ASC-18.

Table 5

DSCS III Technical Data

Transponder	EIRP (dBW)	
X-Band Single Frequency Conversion—6 Channels	Earth Coverage	Narrow Coverage
Chan 1 (MBA)	29.0	40.0
Chan 2 (MBA)	29.0	40.0
Chan 3 (MBA)	23.0	34.0
Chan 4 (MBA)	23.0	34.0
Chan 3 (EC horn)	25.0	
Chan 4 (EC horn)	25.0	
Chan 5 (EC horn)	25.0	
Chan 6 (EC horn)	25.0	
Chan 1, 2 (gimballed dish)	44.0	
Chan 4 (gimballed dish)	37.5	
System Noise Temperature G/T (dB/°K)	Earth Coverage	Narrow Coverage
Multibeam antenna	−16	−1
Earth coverage horn	−14	—
Power output	Chan 1, 2	40 Watts
	Chan 3, 4, 5, 6	10 Watts
Bandwidth	Chan 3, 4	85 MHz
	Chan 2	75 MHz
	Chan 5	60 MHz
	Chan 1, 6	50 MHz
Gain control range	39 dB	
Antenna		
Earth coverage		
MBA (transmit)	Gain 15.0 dB min	
Horn (transmit)	17.0 dB	
Narrow coverage		
MBA (transmit)	Gain 26.0 dB	
	Coverage—1.0° diameter circle on earth disk	
Gimballed dish	Gain 30.2 dB min	
	Coverage—3.0° diameter circle	

SOURCE: GE Astro Space brochure, n.d.

Table 6

DSCS III Transponder Gain Values at 3 dB Backoff

Nominal	1		2		3		4			5	6
TWTA P_O	40 W		40 W		10 W		10 W			10 W	10 W
Downlink Antenna	MBA	GDA	MBA	GDA	MBA	GDA	MBA	GDA	EC	EC	EC
Gain State 1	135.5	134.8	134.0	133.3	128.7	128.9	129.0	128.5	129.2	128.4	129.8
Gain State 2	131.0	130.3	127.5	126.8	122.0	122.2	123.5	123.0	123.7	124.4	123.3
Gain State 3	125.0	124.3	121.0	120.3	115.2	115.4	117.0	116.5	117.2	117.9	117.3
Gain State 6	120.5	119.8	119.0	118.3	113.7	113.9	114.0	113.5	114.2	113.4	117.3
Gain State 4	118.5	117.8	114.5	113.8	109.0	109.2	111.0	110.5	111.2	111.4	110.3
Gain State 7	116.0	115.3	112.5	111.8	107.0	107.2	108.5	108.0	108.5	109.4	108.3
Gain State 5	112.0	111.3	108.0	107.3	102.7	102.9	104.0	103.5	104.2	104.9	104.3
Gain State 8	110.0	109.3	106.0	105.3	100.2	100.4	102.0	101.5	102.2	102.9	100.8
Gain State 9	103.5	102.8	99.5	98.8	94.0	94.2	96.0	95.5	96.2	96.4	94.3
Gain State 10	97.0	96.3	93.0	92.3	87.7	87.9	89.0	88.5	89.2	89.9	89.3

SOURCE: DCA, Circular 800-70-1, Supplement 2, Vol. 2, p. 4-35.

NOTE: Gain States 6 through 10 include the 15 dB pad.

Chapter Eight

LINK FORMULAS

This chapter presents formulas to permit the introduction, one at a time, of the links specified in the scenario. The system parameters that must be specified will be presented as they come into play. Each link can then be evaluated to determine if it is viable. At the same time, the cumulative effect of adding links can be monitored to ensure that the various bandwidth and power constraints are not violated as the satellite is loaded.

A typical diagram of a point-to-point communication satellite circuit is shown in Figure 12. All parameters relating to the uplink bear the subscript u, whether they concern the transmitting earth terminal, the uplink propagation path, or the satellite receiver. Similarly, the downlink parameters bear the subscript d, whether they concern the satellite transmitter, the downlink propagation path, or the receiving earth terminal. All of the parameters required to perform the two link analyses are shown and, with minor obvious differences, are seen to agree with the definitions of the previous chapter.

The basic requirement is to calculate the output EIRPs required at the communication satellite transmitter, (1), and the earth terminal transmitter, (2). In addition, tests for the total EIRP and total signal bandwidth are required at both (1) and (2). Finally, a test for an adequate signal power at the input to the communication satellite receiver, (3), is required.

The formulations will be derived by working backward from the receiving earth terminal through the communication satellite (back-to-front) to the transmitting earth terminal. Note that although the calculated quantities do not bear such a label, they are nonetheless

RAND *MR639-12*

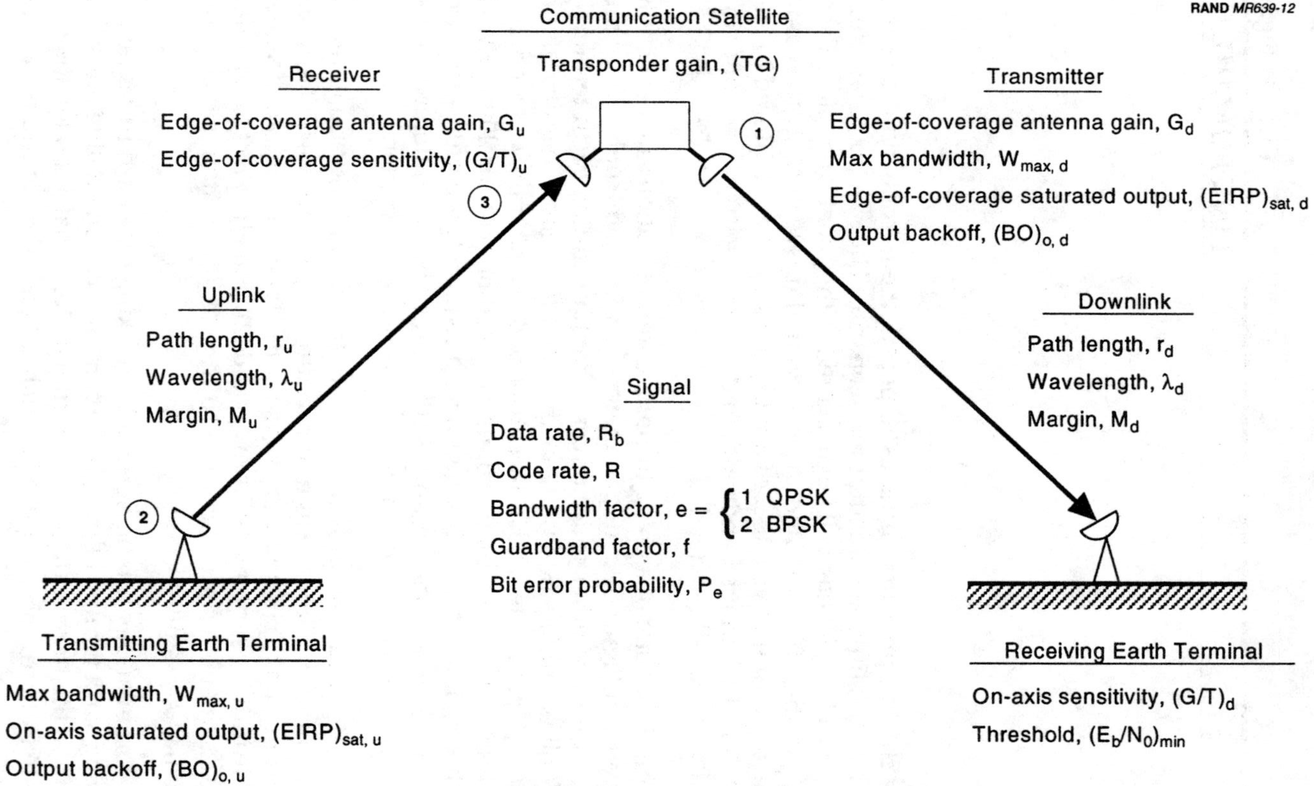

Figure 12—Typical Point-to-Point Communication Satellite Circuit

"required" quantities in the sense that nothing less will close the up- and downlinks while providing the desired link margins.

To calculate the edge-of-coverage downlink EIRP required from the satellite at (1), note that the earth terminal requires an edge-of-coverage received power P_d given by

$$P_d = E_b R_b = N_0\left(\frac{E_b}{N_0}\right) R_b = kT_d\left(\frac{E_b}{N_0}\right) R_b \geq kT_d\left(\frac{E_b}{N_0}\right)_{min} R_b \quad (10)$$

where $k = 1.3806 \times 10^{-23}$ J/K is Boltzmann's constant, T_d is the effective system noise temperature, $(E_b/N_0)_{min}$ is the detection threshold (see Table 3), and R_b is the data rate. To achieve that received power requires an edge-of-coverage power flux density $\Re_d$ at the earth terminal receiving antenna given by

$$\Re_d = \frac{P_d}{A_d} = \frac{4\pi P_d}{\lambda_d^2 G_d} \quad (11)$$

where A_d is the terrestrial receiving antenna's on-axis collecting aperture and G_d is its on-axis gain. The edge-of-coverage downlink EIRP $(EIRP)_d$ required to yield this power flux density is then given by

$$(EIRP)_d = \Re_d\, 4\pi r_d^2 \quad (12)$$

Substituting for P_d and $\Re_d$ from Eqs. (10) and (11), respectively, then yields

$$(EIRP)_d \geq \left(\frac{4\pi r_d}{\lambda_d}\right)^2 \frac{k\left(E_b/N_0\right)_{min} R_b M_d}{\left(G/T\right)_d} \quad (13)$$

which is the first desired result at (1) in Figure 12 and in which $(G/T)_d$ is the on-axis sensitivity of the terrestrial receiver and M_d is the desired downlink margin.

The second result desired at (1) is satisfaction of the downlink EIRP constraint

$$\sum (EIRP)_d \leq \frac{(EIRP)_{sat,d}}{(BO)_{o,d}} \tag{14}$$

where $(EIRP)_{sat,d}$ is the saturated EIRP available from the communication satellite at the edge of coverage and $(BO)_{o,d}$ is the output backoff required according to Table 2.

The third result desired at (1) is satisfaction of the transponder bandwidth constraint

$$\sum W \leq W_{max,d} \tag{15}$$

where the individual signal bandwidths are given by (1) and (2) and $W_{max,d}$ is the bandwidth of the transponder being considered, and where both summations are over all of the signals passing through that transponder.

Ideally, a fully loaded satellite will achieve equality for both constraints. If equality is achieved only in Eq. (14), the transponder is said to be power-limited and some of the available bandwidth remains unused. If equality is achieved only in Eq. (15), the transponder is said to be bandwidth-limited and some of the available power remains unused. Nothing can be done to take advantage of the unused bandwidth (without changing the user mix), but the unused power can be used to increase jam resistance. Accordingly, it will be assumed henceforth that the downlink margin M_d in Eq. (13) will be increased in this case to a level sufficient to ensure that equality is attained in Eq. (14) as well. Of course, this will require that the uplink margin M_u be increased by the same amount.

To calculate the uplink on-axis EIRP required from the earth terminal at (2), note that the power received by the satellite receiver P_u must equal

$$P_u = \frac{(EIRP)_d \, L_i L_o}{(TG) \, G_d} \tag{16}$$

where L_i and L_o denote, respectively, the input and output coupling losses, (TG) denotes the transponder gain as defined by Eq. (9) and listed in Table 6, and G_d denotes the satellite transmitting, i.e., downlink, edge-of-coverage antenna gain. The on-axis power flux density $\Re_u$ required at the satellite receiver then becomes

$$\Re_u = \frac{P_u}{A_u} = \frac{4\pi P_u}{\lambda_u^2 G_u} \tag{17}$$

which parallels Eq. (11) and in which G_u is the satellite uplink receiving antenna gain at the edge of coverage. Finally, the uplink on-axis EIRP $(EIRP)_u$ required becomes

$$(EIRP)_u = \Re_u \, 4\pi r_u^2 M_u \tag{18}$$

or

$$(EIRP)_u = \left(\frac{4\pi r_u}{\lambda_u} \right)^2 \frac{(EIRP)_d \, M_u L_i L_o}{G_u (TG) G_d} \tag{19}$$

which parallels Eq. (13) and yields the first desired result at (2) in Figure 12.

The second result desired at (2) is satisfaction of the EIRP constraint

$$\sum (EIRP)_u \leq \frac{(EIRP)_{sat,u}}{(BO)_{o,u}} \tag{20}$$

where $(EIRP)_{sat,u}$ is the on-axis saturated EIRP available from the earth terminal and $(BO)_{o,u}$ is the output backoff required according to Table 2.

The third result desired at (2) is satisfaction of the earth terminal bandwidth constraint

$$\sum W \leq W_{max,u} \tag{21}$$

where $W_{max,u}$ is the bandwidth of the earth terminal being considered and where both summations are over all of those signals passing through the transponder in question that originate from the earth terminal being considered.

The formulation for $(EIRP)_u$ given by Eq. (19) is based on the detailed specification of the communication satellite elements illustrated in Figure 12. However, some communication satellites, notably those in the INTELSAT series, are specified on an overall basis using specific receiving and transmitting antennas in conjunction with specific transponders. In that case, it is customary to specify the edge-of-coverage uplink power flux density $\mathfrak{R}_{sat,u}$ required at the satellite to produce an edge-of-coverage saturated downlink EIRP $(EIRP)_{sat,d}$. To use these saturation values as surrogates for the quantities L_i, L_o, G_u, (TG), and G_d, which are used to characterize linear operation in Eq. (19), it is necessary to reduce $\mathfrak{R}_{sat,u}$ by the input backoff $(BO)_{lin,i}$, and $(EIRP)_{sat,d}$ by the output backoff $(BO)_{lin,o}$. The values listed in Table 2 for two-carrier operation serve this purpose.

To determine the relationship between the two sets of parameters, substitute $\mathfrak{R}_{sat,u}/(BO)_{lin,i}$ for $\mathfrak{R}_u$ in Eq. (18) and substitute $(EIRP)_{sat,u}/(BO)_{lin,o}$ for $(EIRP)_d$ in Eq. (19). Eliminating $(EIRP)_u$ between Eqs. (18) and (19) then yields

$$\frac{L_i L_O}{G_u (TG) G_d} = \frac{\mathfrak{R}_{sat,u}}{(EIRP)_{sat,d}} \frac{(BO)_{lin,o}}{(BO)_{lin,i}} \frac{\lambda_u^2}{4\pi} \tag{22}$$

Substituting Eq. (22) back into Eq. (19) then leads to

$$(EIRP)_u = 4\pi r_u^2 (EIRP)_d \, M_u \frac{\mathfrak{R}_{sat,u}}{(EIRP)_{sat,d}} \frac{(BO)_{lin,o}}{(BO)_{lin,i}} \tag{23}$$

which is the desired result.

To ensure that the uplink signal is sufficiently strong to allow the uplink noise to be neglected, consider the criterion used in the Aerospace Corporation's COMNET (Lucas et al., 1992), which requires that the ratio of the received carrier power C to the effective system noise power spectral density N_0 at the input to the satellite, i.e., on the uplink, be much larger than the same ratio at the input to the earth terminal receiver, i.e., on the downlink. Using the notation P rather than C leads to the criterion

$$\left(\frac{P}{N_0}\right)_u \geq 10\left(\frac{P}{N_0}\right)_d \tag{24}$$

On the uplink, using Eq. (17) for $\Re_u$, the received input power is given by

$$P_u = \Re_u A_u = \frac{(EIRP)_u}{4\pi r_u^2 M_u}\frac{\lambda_u^2 G_u}{4\pi} = (EIRP)_u\frac{G_u}{M_u}\left(\frac{\lambda_u}{4\pi r_u}\right)^2 \tag{25}$$

so that

$$\left(\frac{P}{N_0}\right)_u = \frac{P_u}{kT_u} = \frac{(EIRP)_u}{kM_u}\left(\frac{\lambda_u}{4\pi r_u}\right)^2\left(\frac{G}{T}\right)_u \tag{26}$$

Similarly,

$$\left(\frac{P}{N_0}\right)_d = \frac{(EIRP)_d}{kM_d}\left(\frac{\lambda_d}{4\pi r_d}\right)^2\left(\frac{G}{T}\right)_d \tag{27}$$

Substituting Eqs. (26) and (27) into Eq. (24) then leads to

$$(EIRP)_u \geq 10\,(EIRP)_d\frac{(G/T)_d}{(G/T)_u}\left(\frac{M_u}{M_d}\right)\left(\frac{r_u}{r_d}\right)^2\left(\frac{\lambda_d}{\lambda_u}\right)^2 \tag{28}$$

If the $(EIRP)_u$ calculated from Eq. (19) fails to satisfy Eq. (28), it must be increased to the equality value. The criterion, Eq. (28), is the desired result at (3) in Figure 12.

Chapter Nine

JAMMING ANALYSIS

The procedure described in Chapter Eight permits a systematic FDM loading of BPSK and QPSK signals on a frequency-translating transponder of a communication satellite. All signals are assigned the same uplink and downlink margins, M_u and M_d, respectively, and the loading proceeds until either the transponder's bandwidth limit or its maximum linear-range power output is approached, whichever comes first. In the first case, the transponder is said to be bandwidth-limited; in the second it is said to be power-limited. The result is the specification, for each signal, of an uplink EIRP required from the originating earth terminal and a downlink EIRP required from the communication satellite.

The above procedure is satisfactory in the absence of jamming, but, inasmuch as the jammer EIRP required to disrupt a given signal can be maximized by using the largest downlink signal EIRP possible, it is clearly unsatisfactory if the loading results in bandwidth-limited transponder operation. This is a consequence, of course, of the fact that not all of the available transponder EIRP is then being used. Therefore, if the transponder loading results in bandwidth-limited operation, it will be assumed, as discussed in Chapter Eight, that the uplink EIRPs will all be proportionally increased to bring the transponder output power to its maximum linear level. Although this could also be done by increasing the transponder gain, that would be self-defeating because both the jamming and desired signals would then be amplified. Because increasing the uplink signal EIRPs results in a corresponding increase in both the uplink and downlink margins, it can be seen that this has the further advantage of providing added protection against propagation disturbances and unintentional interference in the absence of jamming.

Another general factor that must be considered is the location of the jammer relative to the satellite uplink and downlink coverage areas, and the level of sophistication the jammer can exercise. At a minimum, the jammer must be within or, at least, close to the edges of the uplink coverage area so that it can introduce its jamming signal into the transponder it wishes to disrupt. Unless it is also in the satellite's downlink coverage area, the jammer can be at a serious disadvantage, particularly if it does not know the payload configuration being used (see, for example, Figures 6 and 7). A sophisticated jammer can use a spectrum analyzer to examine the received downlink transmission, both to determine the properties of the signals it wishes to jam and to observe its own jamming signal among them. If a jammer is incapable of observing or analyzing the effects of its jamming, either because it is not in the downlink coverage area or because it lacks the necessary receiving equipment, it will be referred to as unsophisticated.

BREAK-LOCK TONE JAMMING

In Appendix D, it is noted that it is characteristic of phase-locked loops of the type used to demodulate BPSK and QPSK signals that they can be caused to break lock if a CW jamming tone that is comparable in power to the desired signal is introduced into the tracking loop. A practical jamming technique is to slowly sweep a CW tone that is slightly more powerful than the desired signal across its carrier frequency. When the tone is within the loop bandwidth, which is typically a few tens of Hertz, the phase-locked loop will break lock with the desired signal in favor of the jamming tone and can remain locked to it as long as the tone remains within the much larger receiver IF bandwidth. Because the tone must not sweep at a rate greater than the square of the loop bandwidth if it is to capture and hold the phase-locked loop, it is limited to a sweep rate of about 1 kHz/sec or so. (Depending on the design of the phase-locked loop, a considerably slower sweep rate might be required.) Thus, the desired signal can be disrupted for considerable periods, even indefinitely, if the jammer is sophisticated and can manipulate the frequency of its CW jamming tone skillfully. Such a jamming strategy can use any configuration, ranging from one large earth terminal radiating a separate jamming tone for each user to be jammed to an ensemble of small earth terminals equal in numbers and comparable in sizes to the individual users to be jammed.

An unsophisticated jammer has a more difficult task. It cannot know the numbers, bandwidths, and center frequencies of the various users and it cannot measure their EIRPs relative to its own jamming tones. At best, it can guess at their numbers and EIRPs and jam with as much EIRP as it can generate using a "comb" of slowly sweeping tones in the hope that it can occasionally or periodically disrupt the desired users. Though admittedly crude, such a technique can prove surprisingly effective against nonpacketized data-transmission systems. A phase-locked loop can regain phase-lock and attain bit and frame synchronism in a fraction of a second when freed from the jammer; the resulting effect on a voice circuit may be negligible, but if such a short break occurs during the reception of a data stream that must be received as a whole, the entire transmission must be reinitiated. Examples are the transmission of facsimile and some databases.

NOISE JAMMING

Although a noise jammer might be expected to use a random waveform with a Gaussian distribution of amplitudes and a uniform power spectral density, this is not usually done in practice. (Such a waveform is often referred to as AWGN, or Additive White Gaussian Noise.) Instead, it is customary to phase-modulate a CW carrier with random noise that has been processed in such a way that the phase-modulated CW carrier has the desired power spectral density (usually uniform over a specified bandwidth). Such a constant-envelope noise signal has two distinct advantages in comparison with AWGN. First, it is more easily and efficiently generated because most amplifiers are peak-power limited and can, therefore, yield a greater output for a constant-envelope noise signal than they can for a signal with a Gaussian (or similar) distribution of amplitudes. Second, as can be seen from Figure B.1, if the jamming signal drives the transponder into saturation, a constant-envelope noise jammer can cause up to 5 dB more small-signal suppression than can an AWGN jammer.

However, the characteristic of noise jamming that is to be exploited here requires that the jamming signal closely resemble AWGN. Hence, if a constant-envelope noise jammer were used against a single phase-locked loop (assuming that the bandwidth of the constant-envelope noise is the same as that of the signal), the following analysis would be inapplicable. Fortunately, this would not be the case in a practical transponder that contained many BPSK or QPSK signals in frequency-division-multiplex, if a single constant-envelope noise jamming signal with a bandwidth equal to that of the transponder were used. In that event, each phase-modulated signal would be received along with only a small segment of the constant-envelope noise jammer's spectrum and Nicholson (1988) shows that such a segment will closely resemble AWGN. Hence, it will be assumed that constant-envelope noise jamming will be employed.

The objective of noise jamming is to increase the receiver noise floor, thereby overcoming the downlink signal margin and driving the received signal-to-noise ratio below threshold. Consider a user whose receiver has an effective (downlink) noise temperature T_d and who is receiving a signal occupying a band W with a downlink margin M_d. To overcome this margin, the jammer must produce at the demodulator a power spectral density P_d^J/W such that

$$kT_d + P_d^J/W = M_d kT_d$$

$$P_d^J/W = (M_d - 1)kT_d \qquad (29)$$

where $k = 1.3806 \times 10^{-23}$ J/K is Boltzmann's constant. The received downlink signal power P_d^S required at the demodulator is given by Eq. (10), which yields, when introduced into Eq. (29),

$$\frac{P_d^J}{W} = \frac{(M_d - 1)P_d^S}{(E_b/N_0)_{min} R_b M_d} \qquad (30)$$

as the required power spectral density for the user in question.

To implement this type of jamming using a constant-envelope noise jammer and make it capable of driving all of the user signals to threshold, it is necessary to provide across the entire transponder bandwidth a jammer power spectral density equal to the maximum value of Eq. (30) calculated over the set of users in the transponder. Then, inasmuch as Eq. (30) can be stated equally in terms of downlink EIRPs, we can write

$$\left(EIRP\right)_d^J = B\left[\frac{\left(EIRP\right)_d^J}{W}\right]_{max} = B\left[\frac{\left(M_d-1\right)\left(EIRP\right)_d^S}{\left(E_b/N_0\right)_{min} R_b M_d}\right]_{max} \tag{31}$$

where B is the transponder bandwidth.

An absolute, rather than a relative, formulation for the required downlink jammer EIRP can be obtained by using Eq. (13) to eliminate the downlink signal EIRP in Eq. (31). This leads to

$$\left(EIRP\right)_d^J = kB\left[\left(\frac{4\pi r_d}{\lambda_d}\right)^2 \frac{\left(M_d-1\right)}{\left(G/T\right)_d}\right]_{max} \tag{32}$$

which illustrates that if the range and downlink margin differences are not great, then it is the most disadvantaged user (i.e., the one with the lowest G/T) that determines the required downlink jammer EIRP. This result also shows a virtually direct relationship between the required jammer EIRP and the downlink margin. That there is no apparent dependence on data rate seems surprising; but recall that the jamming criterion is in terms of a power spectral density. This introduces the signal bandwidth and, hence, indirectly, the data rate.

To gain a feel for the level of jamming required, consider the case of n identical users, for which Eq. (31) reduces to

$$\left(EIRP\right)_d^J = \frac{B}{W}\frac{W\left(M_d-1\right)\left(EIRP\right)_d^S}{\left(E_b/N_0\right)_{min} R_b M_d} \tag{33}$$

Then, noting that B/W = n and using Eqs. (1) and (2) to eliminate W in the numerator yields

$$\left(\mathrm{EIRP}\right)_d^J = \frac{\left(1+2f\right)e\left(M_d-1\right)}{R\left(E_b/N_0\right)_{min} M_d}\left[n\left(\mathrm{EIRP}\right)_d^S\right] \tag{34}$$

where f is the guardband factor, e is the bandwidth occupancy factor, and R is the code rate. As an example, let $f = 0.1$, $e = 2$ (for QPSK from (1), $R = 1/2$ (and, hence, $(E_b/N_0)_{min} = 6.4$ dB from Table 3), and $M_d = 7$ dB. Then from Eq. (34)

$$\left(\mathrm{EIRP}\right)_d^J = 0.880\left[n\left(\mathrm{EIRP}\right)_d^S\right] \tag{35}$$

which shows that the required downlink EIRP is actually 0.56 dB *less* than the total downlink signal EIRP. If BPSK had been used ($e = 1$), the required downlink noise jammer EIRP would have fallen to 3.56 dB less than the total downlink signal EIRP. This seems virtually unbelievable until it is noted that even a minuscule downlink jammer EIRP would suffice, if there were no downlink margin.

This example suggests that the required downlink jammer EIRP, as calculated from Eq. (31), will not exceed the total downlink signal EIRP by much, if at all, unless the users are greatly mismatched. Thus, only in an extreme case would the transponder be driven into saturation in an attempt to provide the required downlink jammer EIRP. It is more likely that a power-limited transponder would be driven to somewhere in its nonlinear region. Unfortunately, it is too difficult to analyze the behavior of an amplifier in this region, so it is not practical to relate the required downlink jammer EIRP precisely to the uplink jammer EIRP required to produce it. However, it can be argued that because of nonlinear effects induced by the jammer, the total interference-to-signal power ratio in the output of a nonlinear device is always equal to or greater than the jam-to-signal ratio in the input.[1] Consequently, Eq. (31) is a conservative statement of the re-

---

[1] If the TWTA were linear, the jam-to-signal ratios at the input and the output would be equal. If, however, the presence of the jammer induces nonlinear effects, the signal power in the output must decrease, because some of the signal power goes into intermodulation products. The jammer power also decreases for the same reason, but the total interference, i.e., jamming plus intermodulation products, then increases.

quired uplink jammer EIRP when put in terms of the uplink signal EIRP. Thus, we can write

$$\left(\text{EIRP}\right)_u^J \leq B\left[\frac{\left(M_d-1\right)\left(\text{EIRP}\right)_u^S}{\left(E_b/N_0\right)_{min} R_b M_d}\right]_{max} \tag{36}$$

where the equality is used to express the upper bound.

An absolute formulation for the uplink jammer EIRP, equivalent to Eq. (32), can be obtained by stating the uplink signal EIRP in terms of its basic parameters. Thus, substituting Eq. (13) into Eq. (19) yields

$$\left(\text{EIRP}\right)_u^S = \left(\frac{4\pi r_u}{\lambda_u}\right)^2 \left[\left(\frac{4\pi r_d}{\lambda_d}\right)^2 \frac{k\left(E_b/N_0\right)_{min} R_b M_d}{\left(G/T\right)_d}\right] \frac{M_u L_i L_o}{G_u\left(TG\right)G_d}$$

When used to eliminate $\left(\text{EIRP}\right)_u^S$ in Eq. (36), this leads to

$$\left(\text{EIRP}\right)_u^J \leq kB\frac{L_i L_o}{G_u\left(TG\right)G_d}\left[\left(\frac{4\pi r_u}{\lambda_u}\right)^2 \left(\frac{4\pi r_u}{\lambda_u}\right)^2 \frac{\left(M_d-1\right)M_u}{\left(G/T\right)_d}\right]_{max} \tag{37}$$

which is the desired result.

BRUTE-FORCE CW JAMMING

This is the simplest of the jamming techniques available against unprotected BPSK and QPSK signals. The jammer need not be sophisticated and needs access only to the uplink. The objective of this type of jamming is to introduce harmful intermodulation products, as discussed in Appendix C, and to drive the communication signals below threshold by power-sharing and small-signal suppression, as discussed in Appendix B. As mentioned, it is assumed that the uplink and downlink signal margins have been increased from their nominal values of 7 dB to a level sufficient to ensure that the transponder is operating at its maximum linear output in the event that the nominal values do not result in power-limited operation.

Two operating regimes of interest result as adding the CW jamming signal drives the transponder out of its linear range. These are the nonlinear region, at low levels of jamming, and the saturation region, at higher levels. Given that the sum of the input signal powers equals the maximum linear input power to the transponder, it can be seen in Figure 11 that input constant-envelope jamming powers less than the input linear backoff $(BO)_{lin,i}$ will carry the transponder into the nonlinear region and that greater input constant-envelope jamming powers will carry it into saturation. In the following, these power levels are referred to in terms of their equivalent uplink EIRPs. Let $(EIRP)_u^{S+\Sigma}$ denote the total EIRP of the uplink signals, where the superscript S denotes the signal of interest and Σ is the sum of all the other signals. Then $(EIRP)_u^{S+\Sigma}$ is the total uplink signal EIRP corresponding to the top of the linear region of operation. Then, $(BO)_{lin,i}(EIRP)_u^{S+\Sigma}$ is the uplink EIRP at which the transponder goes into saturation. Typical values for $(BO)_{lin,i}$ correspond to those listed for the two-carrier case in Table 2.

Nonlinear Operation

The jamming EIRP that confines operation to the nonlinear range is given by

$$ 0 < \left(EIRP\right)_u^J < \left(BO\right)_{lin,i}\left(EIRP\right)_u^{S+\Sigma} $$

The effect of jamming in this case is difficult to analyze, so specific effects will not be presented. Generally speaking, however, it can be said that as the level of jamming is increased, stronger and stronger intermodulation products will be generated and that more and more signal reduction from power-sharing and small signal suppression will be experienced. The consequence will be that all of the signals will experience transmission errors ranging from nil at the low end (where the transponder is still virtually linear) to significant at the high end (where the transponder is virtually hard limiting).

Saturated Operation

The jamming EIRP that brings operation into the saturated region is given by

$$\left(\text{EIRP}\right)_u^J \geq \left(\text{BO}\right)_{\text{lin},i}\left(\text{EIRP}\right)_u^{S+\Sigma}$$

In this case, the transponder can be said to be hard limiting so the results of Appendix A, with respect to power-sharing and small signal suppression, and Appendix C, with respect to intermodulation products, apply. It is impossible to be specific about the effects of intermodulation products because they are so dependent on individual signal EIRPs and bandwidths and center frequencies relative to the frequency of the constant-envelope jammer. In accordance with the discussion in the last paragraph of Appendix C and the illustrative spectra shown in Figure C.3b, it will be assumed that the jammer centers its carrier frequency in the transponder to maximize the presence of intermodulation products within the transponder bandpass.

The relative levels of the various signals and intermodulation products can be inferred from Figure C.2, in which S_1 can be identified with the constant-envelope jammer; S_2 with a signal of interest; the noise N as the sum Σ of the desired signals; and S_{21} as the intermodulation product resulting from the interaction of the signal of interest S_2 with the jammer S_1. Only when the jammer S_1 is larger than the sum Σ of the other signals (i.e., the noise, N) and is also somewhat larger than the signal of interest S_2 (i.e., $S_i \geq 2S_2$) is the intermodulation product S_{21} significant in comparison with S_2 (i.e., $S_{21}/S_2 \geq 0.532$). But this will always be true in the saturated case, because the condition $(\text{EIRP})_u^J \geq (\text{BO})_{\text{lin},i}(\text{EIRP})_u^{S+\Sigma}$ is equivalent to both $S_1 \geq (\text{BO})_{\text{lin},i}N$ and $S_1 \geq (\text{BO})_{\text{lin},i}S_2$ where $(\text{BO})_{\text{lin},i}$ equals 4.5 or 6 dB from Table 2. Hence, every intermodulation product will be very nearly equal in power to the signal that created it, much as depicted in Figure C.3. As a result, all intermodulation products generated by relatively large desired signals that happen to coincide spectrally with relatively weak desired signals can be regarded as disabling them. Even interactions between signals of equal power will probably be disabling to both.

In addition to the above, it is necessary to consider the way in which power-sharing and small signal suppression reduce or overcome the protection afforded by the downlink margin M_d. Consider Eq. (B.9), which gives the ratio of the unjammed to the jammed powers of a given output signal. Assume there are n signals of equal power S in the transponder. Then, in the notation of Eq. (B.9), the low end of saturated operation $(EIRP)_u^J = (BO)_{lin,i}(EIRP)_u^{S+\Sigma}$ is equivalent to $J_{in} = (BO)_{lin,i}(1+n)S_{in}$. In these terms, the condition $(J+\Sigma)_{out} \gg S_{out}$, which is even more true for $(J+\Sigma)_{in} \gg S_{in}$ (because the power-sharing and small signal suppression always favor the jammer), becomes

$$\left(J+\Sigma_{in}\right)_{in} = \left(BO\right)_{lin,i}\left(1+n\right)S_{in} + nS_{in} \gg S_{in} \tag{38}$$

For example, consider a TWTA for which, from Table 2, $(BO)_{lin,i} = 6.0$ dB or (3.98). Also, assume there are n = 10 users. Then, Eq. (38) becomes

$$\left(J+\Sigma\right)_{in} = 43.79\, S_{in} \gg S_{in} \tag{39}$$

which certainly satisfies the condition of Eq. (B.9). To evaluate Eq. (B.9) for these same parameters, note that

$$\gamma = \frac{J}{\Sigma} = \frac{(BO)_{lin,i}(1+n)S_{in}}{n\,S_{in}} = 4.38\ (6.4\ \text{dB})$$

so it is seen from Figure B.1 that the limiter degradation Λ is then about 4.4 dB or (2.75). The form of Eq. (B.9) becomes

$$\frac{S_{out}}{S_{out}^J} = \frac{1+\Lambda\left(\dfrac{J+\Sigma}{S}\right)_{in}}{(BO)_o\left[1+\left(\Sigma/S\right)_{in}\right]} = \frac{1+\Lambda\left[(BO)_{lin,i}\left(1+n\right)+n\right]}{(BO)_{lin,o}\left(1+n\right)} \tag{40}$$

where the right-hand form results when Eq. (38) is used. Then, using $(BO)_{lin,o} = 1.78$ or (2.5 dB) from Table 2, for a TWTA, we have

$$\frac{S_{out}}{S_{out}^{J}} = \frac{1 + 2.75\left[3.98(1+10)+10\right]}{1.78(1+10)} = 7.60 \; (8.81 \text{ dB}) \qquad (41)$$

Thus, an uplink jamming EIRP 6.0 dB greater than the total signal uplink EIRP can drive a TWTA transponder carrying 10 users into saturation and cause 8.8 dB of signal suppression, which exceeds the nominal downlink margin.

From these considerations and calculations, we can conclude that a brute-force CW jammer with an uplink EIRP

$$\left(EIRP\right)_{u}^{J} = \left(BO\right)_{lin,i}\left(EIRP\right)_{u}^{S+\Sigma} \qquad (42)$$

will completely disable every BPSK and QPSK signal in a typical transponder. Note that the CW jammer specified by Eq. (42) is 6 dB larger than the sum of all of the users' signal powers. Hence, it can amount to a sizable jamming terminal. For example, if there were 10 TSC-85 users, each with an EIRP of 70 dB (see Table 4), the jammer would require an EIRP of 86 dBW, which would necessitate, for example, a 20 ft dish (52 dB gain) and a 5 kW amplifier (37 dBW) (assuming 3 dB of losses). In comparison, the TSC-85 has an 8 ft dish and a 500 W amplifier.

This means, therefore, that a brute-force constant-envelope jammer capable of disabling every BPSK and QPSK signal in a typical transponder, though small in comparison with a fixed jammer that approaches the state of the art, is nonetheless one that is not easily transportable or proliferated to the extent that it is physically survivable.

SYSTEM NOISE TEMPERATURE

Several sources of noise contribute to the total noise observed at the output of a receiving system. These include the source noise, the coupling network noise, and the receiver front-end noise.

Source noise accompanies the received signal as it enters the antenna. It may consist of the background (or sky) noise and noise generated by other systems through which the signal has passed before reaching the receiver in question. An example is the noise added to a signal by a communication satellite relay; this noise is usually controlled or made negligible by assuring a more than adequate signal level on the uplink to the communication satellite. On an uplink, background noise will usually dominate and typically be 300 K when a communication satellite antenna is viewing the earth. On a downlink, sky noise may dominate and be as little as a few degrees Kelvin at microwave frequencies under clear sky conditions. However, atmospheric oxygen and water vapor can contribute significant levels of noise at some millimeter waves. Heavy rain can yield high levels of noise at all millimeter waves.

Coupling network noise is added to the received signal as it travels from the receiving antenna to the receiver front end. Its level depends on the temperature of the coupling network and its loss.

Receiver front-end noise is added to the received signal by the low-level input amplifying stages. It is usually expressed in terms of the receiver noise figure NF which is the noise factor in decibels. The noise factor F relates the noise temperature T_R to the reference noise temperature, $T_0 = 290\,\text{K}$, by the expression

$$F = 1 + \left(T_R/T_0\right) \tag{A.1}$$

The noise temperature T_R can range from as little as a few degrees Kelvin for a liquid-helium-cooled maser to 100 K or more for simple uncooled designs. A typical uncooled low-noise amplifier (LNA) may have a noise temperature of 50 K, which corresponds to a noise factor $F = 1.17$ or a noise figure $NF = 0.69$ dB.

To calculate the net effect of these contributions, consider the basic formulation for the network shown in Figure A.1. If there is a coupling network with loss L and a noise temperature T_N following a source with a noise temperature T_S, it can be shown that the noise temperature T_0 at the output of the coupling network (Bedrosian, 1960) is given by

$$T_0 = \frac{T_S}{L} + T_N\left(1 - \frac{1}{L}\right) \tag{A.2}$$

By applying this formula to the receiving system shown in Figure A.2, it can be seen that the overall, or effective, system noise temperature T_0 is given by

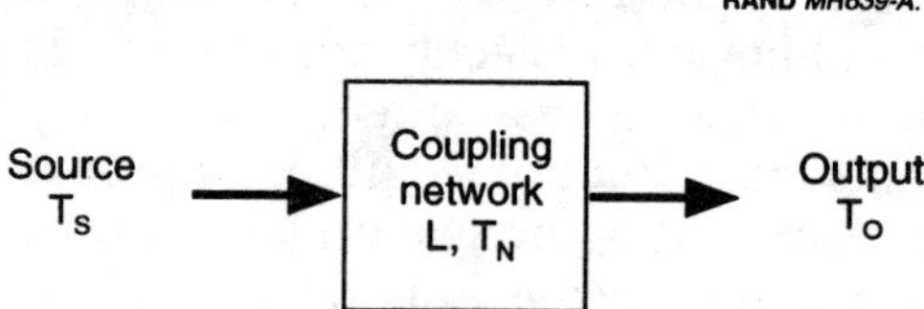

Figure A.1—Generic Coupling Network

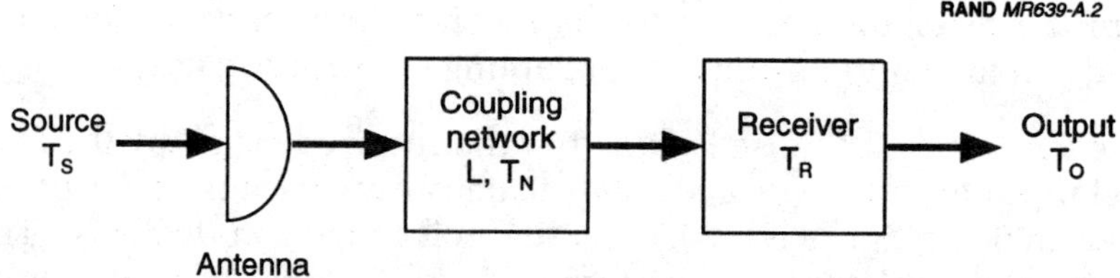

Figure A.2—Generic Receiver Input Network

$$T_0 = \frac{T_s}{L} + T_N\left(1 - \frac{1}{L}\right) + T_R \qquad (A.3)$$

where T_R is the receiver front-end noise from Eq. (A.1). As an example, consider a downlink for which the sky (i.e., the source) temperature is $T_S = 6\,\mathrm{K}$, a coupling network for which $T_N = 300\,\mathrm{K}$ and $L = 3\,\mathrm{dB}\,(2.0)$, and an LNA for which the noise temperature is $T_R = 50\,\mathrm{K}$. Then, $T_0 = 203\,\mathrm{K}$.

POWER SHARING AND SMALL-SIGNAL SUPPRESSION

A transponder in a frequency-translating communication satellite repeater is a bandpass amplifier that has a so-called linear region (see Figure 11) in which its operating characteristic is sufficiently linear that the intermodulation products generated when two or more signals are amplified simultaneously do not significantly affect their individual performances. Therefore, multisignal operation is confined to this region according to the procedures described in the main body of the text.

It is apparent from Figure 11 that the addition of a large interfering signal will cause the transponder to enter the nonlinear region of operation and, in fact, to saturate. This results in three adverse conditions:

1. The level of the intermodulation products will greatly increase. This effect alone can upset the system and prevent normal operation. It will be considered separately in Appendix C and is, therefore, not treated further here.

2. Because the transponder cannot yield a greater output power than its saturation level, the interfering signal obtains its power at the expense of the desired signals. This effect is known as power-sharing and its magnitude approaches the interference-to-signal power ratio when the ratio is large. Its effect, consequently, can be very significant.

3. The desired signals may be affected by a large interfering signal, which could further reduce their output powers because they are

smaller than the interfering signal. This so-called small-signal suppression is not an obvious consequence of using a saturating amplifier. Fortunately, it does not exceed 6 dB in value.

This appendix treats the second and third of these effects.

It is not practical to analyze exactly the nonlinear effects of a saturating amplifier such as that depicted by Figure 11, particularly when the interfering signal is only large enough to cause the transponder to operate in the nonlinear region. Fortunately, the case of an interfering signal large enough to cause saturation, which has been treated, is of great practical interest. In that event, the saturating amplifier is well approximated by a hard limiter, for which an exact analysis is available (Cahn, 1961). Briefly, it can be shown that if the input to a bandpass hard limiter consists of a sinusoid of power S_{in} and an interference of power I_{in} consisting of the sum of a sinusoidal component and a Gaussian noise component, then the ratio of the input to output signal-to-interference power ratios is given by

$$\Lambda = \frac{(S/I)_{in}}{(S/I)_{out}} \tag{B.1}$$

where

$$\frac{1}{\Lambda} = \frac{\pi}{4}(\gamma+1)\left[e^{-\gamma/2}I_0(\gamma/2)\right]^2 \tag{B.2}$$

in which I_0 is a modified Bessel function of the first kind of order zero and γ is the power ratio of the input sinusoidal and Gaussian components of the interference. The quantity $1/\Lambda$, which is plotted in Figure B.1, is usually referred to as the limiter degradation or suppression factor. The analysis is valid when the output interference power is large in comparison with the output signal power—a condition that also validates approximating a saturating amplifier by a hard limiter. It can be seen from Figure B.1 that when the interference is largely noiselike ($\gamma \ll 1$), the signal-to-interference ratio is degraded by about 1.05 dB. When the interference is largely sinusoidal ($\gamma \gg 1$), the signal-to-noise ratio is degraded by as much as 6 dB.

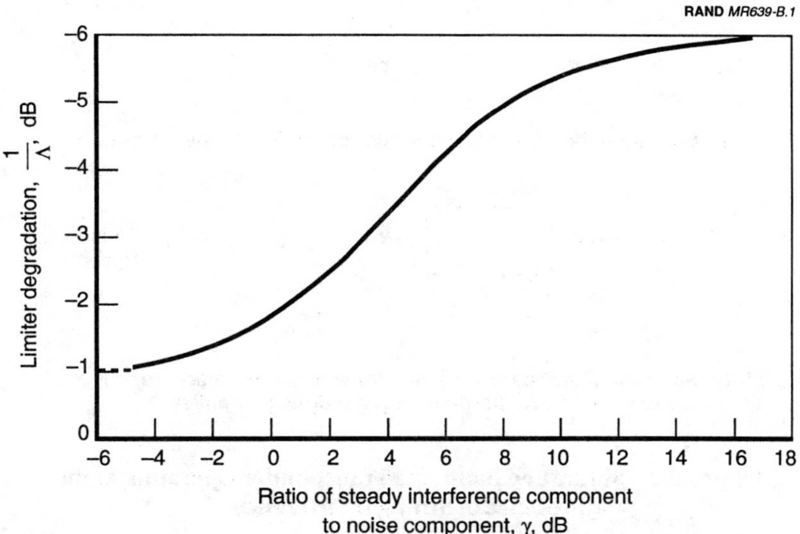

SOURCE: C. R. Cahn, "A Note on Signal-to-Noise Ratio in Band-Pass Limiters," *Trans. IRE*, Vol.IT-7, No.1, January (1961). © 1961 IRE.

Figure B.1—Small-Signal Suppression in a Hard-Limiting Bandpass Amplifier

To apply the foregoing to the case of a large CW (i.e., continuous wave or sinusoidal)[1] input interference to a transponder handling multiple FDM BPSK or QPSK signals, consider Figure B.2a, which depicts normal operation. The input consists of n sinusoidal desired signals, one of which is of interest and has power S_{in} and the remaining n – 1 by their sum $\overset{n-1}{\Sigma}\, S_{in}$. It is assumed in the following that these n – 1 signals are sufficient in number to approximate Gaussian noise when viewed as a sum. In Figure B.2a, in which the transponder is assumed to be operating in its linear range, these signals appear in the output in the same proportion as in the input. In accordance with the discussion of transponder amplifiers in Chapter Seven, this means that there is an output power backoff $(BO)_o$ equal to at least 5 dB plus the appropriate value listed in Table 2.

---

[1]This characterization includes the class of noise jammers in which a CW carrier is modulated by random noise.

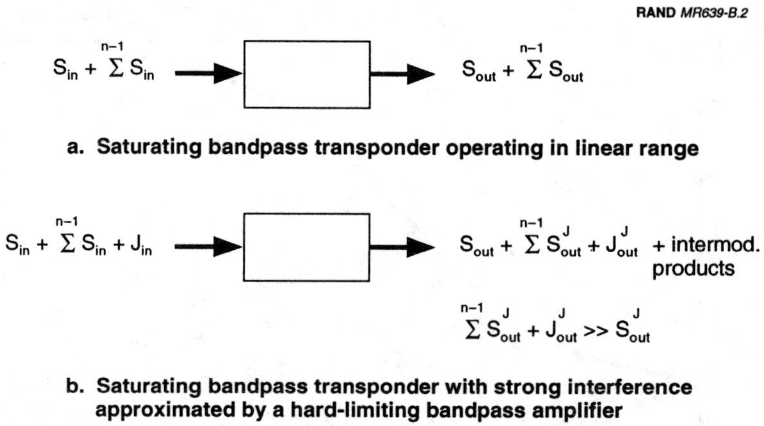

$$S_{in} + \sum^{n-1} S_{in} \longrightarrow \boxed{} \longrightarrow S_{out} + \sum^{n-1} S_{out}$$

a. Saturating bandpass transponder operating in linear range

$$S_{in} + \sum^{n-1} S_{in} + J_{in} \longrightarrow \boxed{} \longrightarrow S_{out} + \sum^{n-1} S_{out}^{J} + J_{out}^{J} + \text{intermod. products}$$

$$\sum^{n-1} S_{out}^{J} + J_{out}^{J} \gg S_{out}^{J}$$

**b. Saturating bandpass transponder with strong interference
approximated by a hard-limiting bandpass amplifier**

**Figure B.2—Saturating Bandpass Transponder Operating in the
Presence of Strong Interference**

A large CW interfering signal denoted by J_{in} is shown added to the desired signals at the input to the transponder in Figure B.2b. To apply Cahn's result, the interference is now considered to be the sum of J_{in} and the $n - 1$ other desired signals (which will be denoted by Σ in the following, for convenience). These signals also appear in the output, together with some intermodulation products, which will be assumed to be negligible in comparison. There is, of course, no backoff in this case because the transponder is now assumed to be saturated. The condition

$$J_{out} + \Sigma_{out}^{J} \gg S_{out}^{J} \tag{B.3}$$

validates the application of Cahn's analysis. Note that the superscript J has been added to S_{out} and Σ_{out} to distinguish them from the corresponding outputs in the linear (i.e., uninterfered with) case.

First, consider the limiter degradation Λ as defined by Eq. (B.1). In the terms of this application,

$$I = J + \Sigma \tag{B.4}$$

Then

$$\Lambda = \frac{(S/I)_{in}}{(S/I)_{out}} = \left(\frac{S_{in}}{J_{in} + \Sigma_{in}}\right) \bigg/ \left(\frac{S^J_{out}}{J_{out} + \Sigma^J_{out}}\right) \tag{B.5}$$

where the parameter γ in Λ is given by

$$\gamma = \left(J/\Sigma\right)_{in} \tag{B.6}$$

Equating the outputs in the two cases depicted in Figure B.2 yields

$$(BO)_o\left(S_{out} + \Sigma_{out}\right) \cong J_{out} + \Sigma^J_{out} + S^J_{out} + IM$$

Dividing through by S^J_{out} and writing

$$\Sigma_{out} = \frac{\Sigma_{out}}{S_{out}}S_{out} = \frac{\Sigma_{in}}{S_{in}}S_{out} \tag{B.7}$$

in the right-hand side leads to

$$(BO)_o \frac{S_{out}}{S^J_{out}}\left(1 + \frac{\Sigma_{in}}{S_{in}}\right) = \left(\frac{J_{out} + \Sigma^J_{out}}{S^J_{out}}\right) + 1 \tag{B.8}$$

(Note the implied linearity between the ratios Σ_{out}/S_{out} and Σ_{in}/S_{in} in Eq. (B.6).) Finally, using Eq. (B.5) in Eq. (B.8) to introduce Λ yields

$$\frac{S_{out}}{S^J_{out}} = \frac{\Lambda\bigg/\left(\frac{S_{in}}{J_{in} + \Sigma_{in}}\right) + 1}{(BO)_o\left(1 + \Sigma_{in}/S_{in}\right)}$$

or

$$\frac{S_{out}}{S_{out}^{J}} = \frac{1 + \Lambda\left(\dfrac{J+\Sigma}{S}\right)_{in}}{(BO)_{o}\left[1 + \left(\Sigma/S\right)_{in}\right]} \qquad \left(J+\Sigma\right)_{out} \gg S_{out} \qquad (B.9)$$

which, together with Eq. (B.7), is the desired result in that it shows the effect of a steady interference on one of a number of steady FDMA signals.

INTERMODULATION PRODUCTS

As mentioned in Appendix B, passing two or more narrowband signals through a nonlinear device produces intermodulation (or cross) products that appear in the output as additional interfering signals. The calculation of these intermodulation products is a complex process with useful analytical results available only for the bandpass hard limiter and an error function type of the smooth bandpass limiter. Numerical results are available only for the hard limiter in the cases of two signals of arbitrary amplitude in the presence of noise (Jones, 1963), three or more signals of equal amplitude in the absence of noise (Shaft, 1965); and signals of equal amplitude in the presence of noise (Shaft, 1965).

Inasmuch as the interest here is the generation of intermodulation products by an interfering signal that is large in comparison with the desired signal, attention will be restricted to the first of these references. The restriction to two signals is unfortunate because the application of interest is the case of one interfering signal and a number of desired signals, often having different amplitudes, that are small in comparison with the interfering signal. However, it will be seen that if it is assumed that the consequences of having more than one desired signal are to make matters worse (by adding still more intermodulation products), then the result can become unacceptable when there are two or more desired signals.

The numerical results from Jones (1963) are presented in Figures C.1 and C.2. Though not directly relevant to the issue of intermodulation products, the plot shown in Figure C.1 is included because it gives further insight into the matter of small signal suppression, which

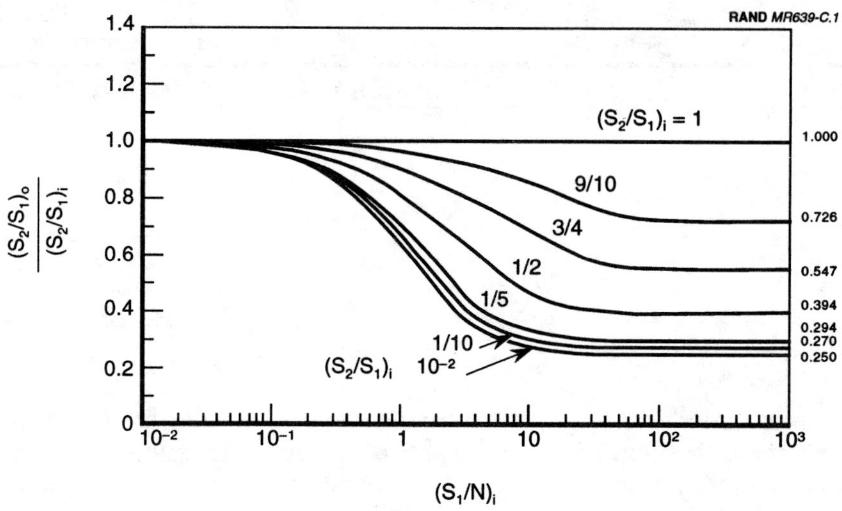

SOURCE: J.J. Jones, "Hard Limiting of Two Signals in Random Noise," *Trans. 1EEE*, Vol.IT-9, No.1, January (1963). © 1963 1EEE.

Figure C.1—The Ratio of the Output Signal-to-Signal Power Ratio to the Input Signal-to-Signal Power Ratio as a Function of the Larger Input Signal-to-Noise Ratio

was treated in Appendix B and was presented as the limiter degradation plot of Figure B.1. In Figure C.1, the input signal S_{1_i} can be regarded as the interference and S_{2_i} can be regarded as the desired signal because the plot parameter is the ratio $(S_2/S_1)_i \leq 1$. It can be seen that the ratio of these two signals at the output $(S_2/S_1)_o$ to their ratio at the input $(S_2/S_1)_i$ is unchanged when the interference-to-noise ratio at the input is very small $(S_1/N)_i \ll 1$, regardless of the signal ratios at the input.[1] That is, the noise appears to have a

---

[1] This is not in contradiction with Figure B.1, which shows an 11 dB degradation when the steady component of the interference is small in comparison with the noise component. In that analysis, the interference is the sum of the steady and noise components, so holding the interference-to-signal ratio constant as the ratio of the steady to the noise component decreases is equivalent to having the steady component vanish in the limit. Here, the interference consists only of a steady component whose amplitude is fixed. Thus, in this case the noise tends to infinity in the limit as the interference-to-noise ratio decreases.

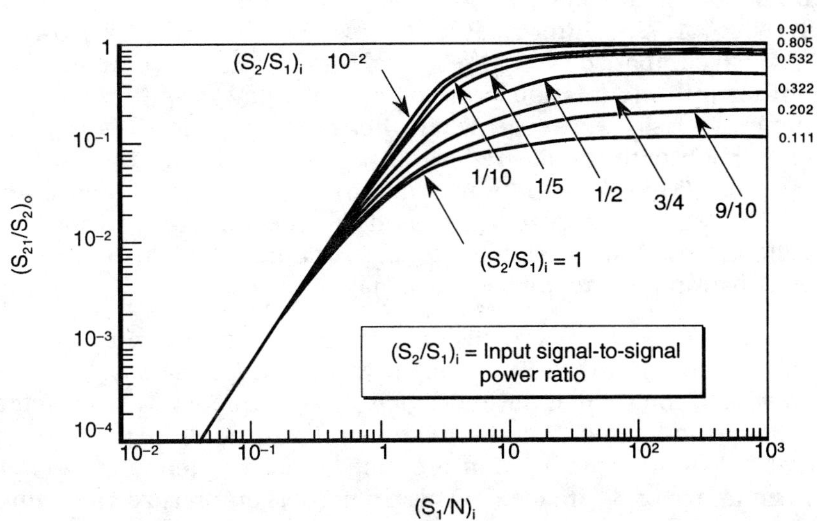

SOURCE: J.J. Jones, "Hard Limiting of Two Signals in Random Noise," *Trans. 1EEE*, Vol.IT-9, No.1, January (1963). © 1963 1EEE.

Figure C.2—The Power Ratio of the Strongest Intermodulation Product to Weakest Output Signal as a Function of the Larger Input Signal-to-Noise Ratio

mitigating effect. Though interesting, this case is of little practical interest.

When the interference-to-noise ratio at the input is very large, $(S_1/N)_i \gg 1$, the small-signal suppression is evidenced as a function of the signal ratios at the input. When the desired signal is very small in comparison with the interference $(S_2/S_1)_i \ll 1$, it displays the 6 dB small-signal suppression also shown in Figure B.1.

The ratio of the strongest output intermodulation product S_{21_o} to the desired output signal S_{2_o} is plotted as a function of the input interference-to-noise ratio $(S_1/N)_i$ for various values of the input desired-to-interfering signal ratio $(S_2/S_1)_i$. The subscript notation specifies the frequency of the component. Thus, S_1 is at frequency, f_1, and S_2 is at frequency, f_2. The strongest intermodulation product S_{21} is then at the frequency $2f_1-f_2$. When the input signals are of equal amplitude $(S_2/S_1)_i = 1$ and the noise is negligible, $(S_1/N)_i \gg 1$, this intermodulation product is down by 9.54 dB (0.111) in comparison with the desired signal (and is accompanied by another intermodulation product S_{12} of equal amplitude at frequency $|f_1-2f_2|$). However, when the interfering signal is large in comparison with the desired signal $(S_2/S_1)_i \gg 1$, and the noise is negligible $(S_1/N)_i \gg 1$, the intermodulation product S_{12} vanishes and S_{21} equals the desired signal in the output. This is a startling result because it shows that the output contains an extraneous signal equal in amplitude to the desired signal.

The output spectrum for this case is illustrated in Figure C.3a in which the interference S_1 at frequency f_1 is depicted as a pure sinusoid and the smaller desired signal S_2 at frequency f_2 is depicted as being modulated, but still of constant amplitude (e.g., BPSK or QPSK). The intermodulation product S_{21} at frequency $2f_1-f_2$ is a mirror image of S_2 about S_1. This property is emphasized in Figure C.3a by placing arrows within the modulation spectral envelopes. We now speculate (without proof, but with considerable intuition) that were there another desired signal S_2' at frequency f_2' and if the input interference-to-total-signal ratio were sufficiently large, i.e., $(S_1/(S_2+S_2'))_i \gg 1$, then there would appear in the output a similar reversed intermodulation product S_{21}' at frequency $2f_1-f_2'$. That is, a sort of superposition principle would apply. This is illustrated in Figure C.3a for a second desired signal S_2' at a frequency f_2' greater than, i.e., to the right of, S_2.

The consequence of interest in the current application occurs when the interfering signal S_1 is at a frequency lying between the two desired signals. Then, as illustrated in Figure C.3b, each one's strongest intermodulation product can fall wholly or partly on the other signal. Considering that all of the desired signals and intermodulational products may be of the same relative size, the mutual interference will become intolerable.

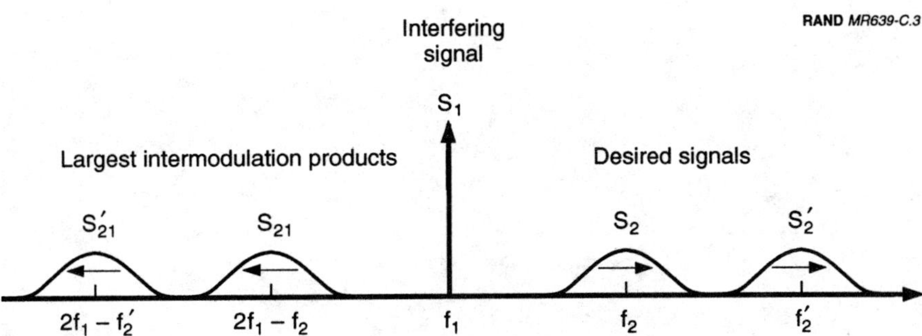

a. Interfering signal outside desired signals

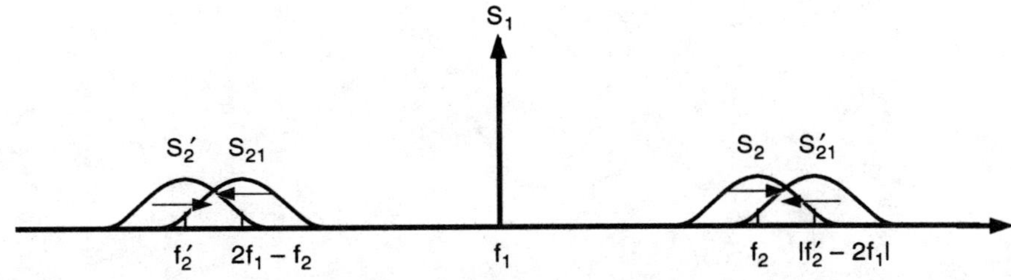

b. Interfering signal between desired signals

**Figure C.3—Spectra of Desired Signals and Their Strongest
Intermodulation Products**

INTERFERENCE IN PHASE-LOCKED LOOPS

All commercial and many military communication satellite data transmission systems use conventional BPSK or QPSK signals in frequency division multiplex. Such signals are not protected against jamming or, even, strong unintentional interference. This appendix treats the behavior of such circuits in the presence of deliberate, steady (i.e., CW) interference. The results will not stem from rigorous theoretical analyses. Instead, they will reflect the best judgment of researchers in the field as derived from experience and some limited experimental evidence.

A phase-locked loop uses a feedback circuit that permits the local oscillator in the receiver to achieve and maintain phase-lock with the carrier of the modulated signal. This is done with an effective loop bandwidth that is invariably much narrower than the bandwidth of the modulated signal—values of a few Hertz to a few tens of Hertz being common. Inasmuch as the signal-to-noise ratio required in the signal bandwidth, i.e., the threshold for adequate detection, has a typical value of a few dB, it can be seen that the signal-to-noise ratio in the loop bandwidth will be much larger—values of 30 dB or more being common. As a result, the tracking loop is more robust than the receiver's demodulation ability. That is, even though the error rate becomes unacceptable for signal levels just below threshold, much larger decreases in signal level will be required before the tracking loop is no longer able to retain phase-lock. Measurements on a receiver having a loop bandwidth of 12 Hz showed no effect at a signal-to-noise ratio of 7 dB, noticeable cycle slipping at 5 dB, and loss of lock at 3.5 dB (Britt and Palmer, 1967).

The loop signal-to-noise ratio can be reduced by a natural cause, such as increased path attenuation, which is one reason for providing a margin of power in the link design. Another cause is CW interference lying within the loop bandwidth. This was demonstrated analytically in Yoon and Lindsey (1982), in which an expression for the probability density function was derived for the case of a steady sinusoid at the carrier frequency. Calculations illustrated the growth of the spread of the loop phase error as the interference level was increased. The authors did not reach quantitative conclusions, but the numerical example showed a clearly unacceptable spread at a signal-to-interference ratio of 8 dB.

Measurements for a stationary or slowly sweeping CW interference within the loop bandwidth are presented in Britt and Palmer (1967). Effects were negligible if the frequency of the interfering signal was well outside the loop bandwidth or if the interfering signal was being swept too rapidly. A nominal maximum sweep rate would be to traverse the loop bandwidth in no more than its time constant (i.e., the reciprocal of the loop bandwidth). This gives a maximum sweep rate equal to the square of the loop bandwidth.

At high loop signal-to-noise ratios, the loop mean-square error increased by about 3 dB for each 6 dB increase in interference-to-signal ratio. Cycle slipping began when the interference-to-signal ratio exceeded –3 dB. Interfering signals as much as 18 dB below the desired signal caused a measurable degradation in the loop mean-square error, which is consistent with the calculations of Yoon and Lindsey (1982).

At low loop signal-to-noise ratios (6 to 12 dB), the interference caused an increase in the loop mean-square error of about 1 dB at an interference-to-noise ratio of –3 dB. In the absence of interference, cycle slipping or loss of lock was observed at loop signal-to-noise ratios of 6 to 9 dB. *Interference levels equal to the signal level consistently caused loss of lock, regardless of the loop signal-to-noise ratio.*

CW interference within the signal bandwidth but not in the loop bandwidth is a less severe threat. Blanchard (1974) presented theoretical and experimental results; the theory was extended by Levitt (1981), who also presented some numerical results. Both authors derived a limited condition for loss of lock, showing that the signal-to-interference ratio required to break lock goes as the square of the frequency by which the interfering signal lies outside the loop bandwidth. Recalling that an interference-to-signal ratio of 0 dB suffices to break lock when the interfering signal is within the loop bandwidth, it can be appreciated that considerably higher levels are required to break lock when the interfering signal is outside the loop bandwidth.

It is emphasized that the numerical values cited above should not be taken as absolutes. Each case should be considered individually if specific values are desired. However, the trends are probably representative and can be used with confidence for comparative purposes. The criterion that a slowly sweeping interfering signal within the loop bandwidth can cause loss of lock when it is comparable to the signal level, however, can be used with some confidence in the general case. It is true that smaller interfering signals within the loop bandwidth will degrade performance and that larger interfering signals outside the loop bandwidth can cause loss of lock, but this information is less useful.

To apply the foregoing to the case of a link relayed by a frequency translating transponder in a communication satellite, it is necessary to note that the interfering signal must be introduced on the uplink. If the interfering signal is comparable in power to the desired signal level, the transponder will likely continue to operate in, or just barely out of, the linear region, with the result that the foregoing will apply in much the fashion in which it was presented. Stated in terms of a slowly sweeping CW interfering signal that lies within the loop bandwidth of a desired signal, it can be concluded that if the interfering signal is comparable in level with the desired signal, then it will cause loss of lock, thereby disabling the link. Thus, an entire transponder can be disrupted by a set of interfering transmitters equal in number and size to those radiating the desired signals. Note that this will increase the total power on the uplink by only 3 dB, which is less than the input backoff of the satellites' power amplifier (see Table 2). As a result, the amplifier will not even be driven to saturation. Note also that no existing or planned satellite has the capability of detecting the presence, much less the location, of such sources of interference. Hence, their effect cannot be negated by antenna nulling.

If the interfering signal is much stronger than all of the desired signals, then the effects of intermodulation, product generation, power-sharing, and small-signal suppression discussed in Appendixes B and C come into play. Such an interfering signal might, by chance, lie within the signal passband of one of the desired signals, which, of course, would be overwhelmed. The combination of signal strength reduction because of power-sharing and small-signal suppression together with the interference caused by the numerous intermodulation products would disable all of the desired signals if the interference could drive the transponder into hard limiting. For a moderate number of desired signals in a transponder, this tactic would require considerably more power in the single large interferer than in the sum of the small interferers just considered, but it would be considerably simpler to do. However, it would be a poor tactic against a satellite such as DSCS III, which can detect the presence of such an interferer, determine its direction, and place upon it a null from its multibeam antenna.

Bedrosian, E., "On the Noise Temperature of Coupling Networks," *Trans. IRE,* Vol. MTT-8, No. 4, July 1960, p. 463.

Blanchard, A., "Interferences in Phased-Lock Loops," *Trans. IEEE,* Vol. AES-10, No. 5, September 1974, pp. 686–697.

Britt, C. L., and D. F. Palmer, "Effects of CW Interference on Narrow-Band Second Order Phase-Lock Loops," *Trans. IEEE,* Vol. AES-3, No. 1, January 1967, pp. 123–134.

Cahn, C. R., "A Note on Signal-to-Noise Ratio in Band-Pass Limiters," *Trans. IRE,* Vol. IT-7, No. 1, January 1961, pp. 39–43.

Clark, George C., J. Bibb Cain, *Error-Correction Coding for Digital Communications,* Plenum Press, New York, 1981.

Haccoun, David and Guy Bégin, "High-Rate Punctured Convolutional Codes for Viterbi and Sequential Decoding," *IEEE Transactions on Communications,* Vol. 37, No. 11, November 1989, pp. 1113–1125.

INTELSAT Earth Station Standards, Document IESS-410 (Rev. 1), 13 June 1990.

Jones, J. J., "Hard Limiting of Two Signals in Random Noise," *Trans. IEEE,* Vol. IT-9, No. 1, January 1963, pp. 34–42.

Levitt, B. K., "Carrier Tracking Loop Performance in the Presence of Strong CW Interference," *Trans. IEEE,* Vol. COM-29, No. 6, June 1981, pp. 911–916.

Lindsey, William C., and Marvin K. Simon, *Telecommunication Systems Engineering*, Prentice-Hall, Inc., Englewood Cliffs, New Jersey, 1973.

Lucas, Richard H., James K. Young, and Russel D. Raymond, *COMNET, Commercial Network Exploration Tool Users Manual*, The Aerospace Corporation, 1992.

Nicholson, David L., *Spread Spectrum Signal Design*, Computer Science Press, Rockville, Maryland, 1988.

Shaft, Paul D., "Limiting of Several Signals and Its Effect on Communication System Performance," *Trans. IEEE*, Vol. COM-13, No. 4, December 1965, pp. 504–512.

U.S. Department of Commerce, "Tables of Frequency Allocations and Other Extracts: Manual of Regulation and Procedures for Federal Radio Frequency Management," National Telecommunications and Information Administration, September 1989.

Yoon, C. Y., and W. C. Lindsey, "Phase-Locked Loop Performance in the Presence of CW Interference and Additive Noise," *Trans. IEEE*, Vol. COM-30, No. 10, October 1982, pp. 2305–2311.